WHAT I REQUIRE FROM LIFE

J.B.S. Haldane, 1963, near lake Winona, in Madison, Wisconsin by Dr Klaus Patau, Department of Medical Genetics, University of Wisconsin

WHAT I REQUIRE FROM LIFE

Writings on Science and Life from J. B. S. Haldane

EDITED BY

KRISHNA DRONAMRAJU

OXFORD
UNIVERSITY PRESS

OXFORD
UNIVERSITY PRESS

Great Clarendon Street, Oxford OX2 6DP

Oxford University Press is a department of the University of Oxford.
It furthers the University's objective of excellence in research, scholarship,
and education by publishing worldwide in

Oxford New York

Auckland Cape Town Dar es Salaam Hong Kong Karachi
Kuala Lumpur Madrid Melbourne Mexico City Nairobi
New Delhi Shanghai Taipei Toronto

With offices in

Argentina Austria Brazil Chile Czech Republic France Greece
Guatemala Hungary Italy Japan Poland Portugal Singapore
South Korea Switzerland Thailand Turkey Ukraine Vietnam

Published in the United States
by Oxford University Press Inc., New York

First published 2009

British Library Cataloguing in Publication Data
Data available

Library of Congress Cataloging-in-Publication Data
Data available

Typeset by SPI Publisher Services, Pondicherry, India
Printed in Great Britain
on acid-free paper by
Clays Ltd., St Ives plc

ISBN 978–0–19–923770–8

1 3 5 7 9 10 8 6 4 2

CONTENTS

PART I Essays from the Marxist period (1937–50)

PART II Essays from the Indian Period (1957–64)

FOREWORD

by Sir Arthur C. Clarke, CBE

J. B. S. Haldane was perhaps the most brilliant science popularizer of his generation. Starting in 1923 with *Daedalus; or Science and the Future*, he must have delighted and instructed millions of readers. Unlike his equally famous contemporaries, Jeans and Eddington, he covered a vast range of subjects. Biology, astronomy, physiology, military affairs, mathematics, theology, philosophy, literature, politics—he tackled them all. He also wrote a workmanlike *novella, The Gold Makers*, and a charming tale for children, *My Friend Mr. Leakey*.

Haldane's scores of essays may still be read with great profit. They appeared in such varied places as *Harper's Magazine*, the *Saturday Evening Post*, the *Strand Magazine, The Spectator*, the *Daily Express*—and, of course, the *Daily Worker*.

Haldane wrote a series of articles for the *Daily Worker* from 1937 to 1949, at a time when he subscribed to the ideals of socialism and communism (which, in the real spirit of open-mindedness, he changed later). This was followed by regular contributions to Indian newspapers during the last few years of his life (1957–64). Written in his usual lucid style, their breadth and depth of coverage indicate that he was truly a multi-disciplinary scientist—an effective bridge between the 'two cultures'.

It is a pity that these essays are not as widely known today as they ought to be. So I am very glad that one of Haldane's former students and long-standing associates, Professor Krishna Dronamraju, is bringing out this collection. Krishna has previously edited or compiled a number of publications based on the work of his mentor. We can derive fresh insights from this collection of works by one of the greatest science communicators of our times.

I was first attracted to Haldane's writings by the element of extrapolation they contained. He obviously was sympathetic to science fiction and astronautics; indeed, this paragraph was contained in his very first book, *Daedalus*:

> I should have liked, had time allowed, to have added my quota to the speculations which have been made with regard to inter-planetary communication. Whether this is possible I can form no conjecture; that it will be attempted, I have no doubt whatever.

It was through space flight that I made my first encounter with Haldane. In 1951, as Chairman of the British Interplanetary Society, I invited him to give our organization a paper on the biological aspects of space flight. Although it was arranged at very short notice, the lecture was a great success. He dealt with three problems: how humans would live in spaceships, how they would live on other planets, and what sort of life they might find there. At that time—some six years before the Space Age dawned—these were not subjects with which many reputable scientists cared to be associated.

Haldane's 1951 paper still contains some interesting ideas. He must have been one of the first to point out the dangers of solar flares and to suggest that space voyages should be made during periods of minimum solar activity. And, with his tongue firmly in his cheek, he suggested that we should take seriously the hypothesis that life has a supernatural origin—from which he concluded that, as there are 400,000 species of beetles on this planet, but only 8,000 species of higher animal forms, 'the Creator, if he exists, has a special preference for beetles, and so we might be more likely to meet them than any other type of animal on a planet that would support life.'

The second time our paths crossed was in November 1960, by which time both Haldane and I had settled down in the East (him in Bhubaneswar, eastern India, and myself in Colombo, Ceylon).

The Ceylon Association for the Advancement of Science invited Haldane to address its annual meeting, and he arrived with Krishna Dronamraju and another colleague. Over the next few days, I had the stimulating experience of showing them around Colombo, and having them over at my house.

One evening after dinner, we screened the movie *Beneath the Seas of Ceylon*, which my diving partner Mike Wilson and I had recently made—the first time the Indian Ocean's undersea wonders were captured on film. I can still remember how Haldane watched it with such delight, showing the sense of wonder that is the hallmark of the great scientist.

More details of this second encounter are found in my essay 'Haldane and Space', which I wrote at Krishna's invitation for the excellent memorial volume he edited, *Haldane and Modern Biology*. It is also included in my own collection of essays, *Report on Planet Three and Other Speculations*.

Haldane and I never met again, but we sustained a correspondence. His letters, usually handwritten and often running into a thousand words, were so full of ideas as his agile mind jumped from one subject to another that they were both good fun and hard reading.

In 1962, shortly after I had won the UNESCO Kalinga Prize for the popularization of science, I received an invitation to stay with the Haldanes. It opened with this rather ambiguous compliment:

> Personally I should also like to see you awarded a prize for theology, as you are one of the very few living persons who has written anything original about God. You have, in fact, written several mutually incompatible things.... if you had stuck to one theological hypothesis, you might be a serious public danger.

To my lasting regret, I was unable to accept Haldane's hospitality due to an unexpected illness (polio) and other distractions. Shortly afterwards, Haldane himself succumbed to cancer—but not before he had made fun of his illness in a witty poem called 'Cancer is a funny thing'. He certainly did not go gentle into that good night...

Two years later, Carl Sagan and Josef Shklovski dedicated their pioneering book *Intelligent Life in the Universe* (Holden-Day, 1966), as follows:

> To the memory of John Burdon Sanderson Haldane, FRS, member of the National Academies of Science of the United States and of the Soviet Union, member of the Order of the Dolphins, and a local example of what this book is about.

You are about to sample the products of one of the finest intellects it was ever my privilege to know.

Sir Arthur C. Clarke
Fellow, King's College, London
Colombo, Sri Lanka

Sir Arthur C. Clarke passed away on March 18, 2008, while this book was in press.

PREFACE

Haldane, Marxism, and Popular Science

by James F. Crow
University of Wisconsin

Before his Communist period, which started in the 1930s, J. B. S. Haldane was already widely known, both as a scientist and as a writer. His scientific credentials were well established by two greatly admired books, '*The Causes of Evolution*' and '*Enzymes*', in addition to well over 100 scientific papers. His popular books included '*Heredity and Politics*' and '*The Marxist Philosophy and the Sciences*'. He also wrote science fiction, '*Daedalus, or Science of the Future*', '*Possible Worlds*', and a children's book, '*My Friend Mr. Leakey*'. And there were others.

As a popular writer, Haldane was in the same league as Bertrand Russell. Both could explain complex ideas with clarity; both wrote with a style that was simple, but not simplistic; both could be witty; both were willing to take on unpopular causes; and both invited criticism because their ideas were so clear. It is easier to criticize what you understand. Haldane's first wife, Charlotte, was a journalist. She honed his writing skills and helped him tailor his work to the market. As a result he published in such contrasting places as the *Atlantic Monthly*, *Harper's Magazine*, the *Saturday Evening Post*, the *Nation*, various newspapers, and the communist *Daily Worker*. His popular writing enhanced both his reputation and his income.

Haldane's writing had remarkable clarity, no matter what audience he was addressing, and was suited to that audience. His '*The Causes of Evolution*' is often compared with R. A. Fisher's great book '*The Genetical*

Theory of Natural Selection.' The two could hardly be more different. Haldane's is easy to read, even in the mathematical sections. His sentences are short and direct. His examples are apt and illuminating. As a result, the book was instantly popular and can still be read with profit. Fisher's book was deeper. The sentences are long and convoluted. The writing is elegant, but often obscure, and these obscurities are still being debated. People often ask, what did Fisher *really* mean? Although both Fisher and Haldane were gifted in mathematics, Fisher was more creative. One thinks that with Haldane's work, one could duplicate it, given sufficient time. Fisher's ideas, in contrast, seem to come out of nowhere.

In addition to both being statistical outliers, they had another trait in common: both loved to argue. Especially, they loved to argue with each other. When they were both at the Rothamsted Experiment Station, teatime was regularly enlivened by a Haldane–Fisher debate, often interminable and on any subject whatsoever.

Haldane's colourful lifestyle appeared even in his technical writing. He pioneered, with the Russian scientist A. I. Oparin, the idea that the earliest forms of life could be simple because so much organic chemistry had happened in the environment prior to life, freeing them from having to make the compounds for themselves. His expression that the original forms of life would find themselves in an environment of 'hot dilute soup' is one of the most famous *bon mots* in all biology.

In 1942, Haldane wrote a semi-technical book, '*New Paths in Genetics*', based on a series of public lectures given in Holland in 1940. He later wondered how many of his audience perished or ended up in concentration camps. I read the book when it was new and, even though I knew a great deal of Haldane's work, I was enormously impressed. One small innovation, which instantly invaded the whole field of genetics, was his translocating the words, *cis* and *trans*, from chemistry to genetics, replacing the awkward and misleading *coupling* and *repulsion* for designating the two linkage phases. The expressions are now in everyday use, and most geneticists are unaware of their source. I saw for the first time the chemical steps in phenyalanine metabolism, in which specific deficiencies led to several inborn errors of metabolism. Haldane presented the complete pedigree of Queen Victoria and her descendants. The good queen inflicted enormous damage on the royal families of Europe. The British habit of exporting daughters while keeping the sons at home had a devastating effect. This was historically most important in the family of Tsar Nicholas of Russia, who had married one of Victoria's granddaughters. The tsarina's preoccupation with the suffering of young Tsarevitch Alexis may have been a contributing factor to the downfall of the Romanoffs.

Haldane also described a number of pedigrees showing linkage between the X-chromosomal genes for hemophilia and colour-blindness, and for the first time estimated the map distance between the loci of the two genes. He also measured linkage between colour blindness and night blindness, an analysis notable more for its complexity than for its importance. The algebra was awesome. In Haldane's words, 'This reduces to an algebraic equation of the twenty-second degree with integral coefficients, and I think I am probably safe in saying that this is the first occasion on which such an equation has arisen in the course of scientific work.' He was able to find a numerical solution. This is all the more remarkable in that this was before computers and he did the calculations by hand. I am sure he enjoyed it; one of Haldane's eccentricities was a love of doing complicated arithmetic and algebraic calculations. Despite the complexity of the subject, his writing is remarkably lucid and concise—to me, this book is Haldane at his best.

In the 1920s and early '30s Haldane was critical of materialism. At least two things turned him leftward in the mid-1930s. One was his wife, Charlotte, who was a Communist. Another was the Spanish Civil War. In 1936 he got involved on the side of the Loyalists. There he began his writing for the communist *Daily Worker*. Returning to London, but not abandoning his activity on behalf of the Spanish Loyalists, in 1937 he became the chairman of the Editorial Board of *Daily Worker* as well as its science correspondent. This continued until his last article in 1950.

For several years he wrote a weekly article on science, a number of which are included in this volume. But this is only a tiny sample, since he contributed 345 articles to this source. And that is not all. He wrote hundreds of articles for other journals in addition to book reviews. Many of these articles were written on the train as he commuted to and from work. Haldane always carried a note pad and, wherever he was, he would continue writing, taking up where he had earlier left off. He wrote his popular articles this way, and in the same notepad, turned upside down, he continued with the calculations that he so enjoyed. Haldane had a near-perfect memory, so he did not need to consult references as he wrote—a great saving of time. Yet, even taking all this into account, the total amount of his output is still hard to believe.

Although during the war I was aware of Haldane's activities, I never saw any of his war-time popular writings until later. In 1947, he collected some of these articles into a book entitled '*What is Life*'. As well acquainted as I was with Haldane's erudition, these still astounded me. This shows up on page after page. So does his facility of expression, and so does his willingness to shock. A year earlier, he had assembled another set, published as 'A Banned Broadcast'. And a year later still came another collection, '*Science* Advances'.

One article, reprinted here, was 'How to write a popular scientific article.' It has lots of commonsense advice: know a great deal about the subject, much more than you can put in this article; consider for whom you are writing; start with a known fact 'a bomb explosion, a bird's song, or a cheese'; from this proceed to a scientific principle; try to include some news, presumably not already known to the reader; use active verbs rather than passive; keep the sentences simple and short; finally, show it to a friend, 'if possible a fairly ignorant one'. He discusses, as an example, a possible article about cheese manufacture. 'If you are writing for a highbrow journal you may quote the passages on cheese from Euripides' *Cyclops*, if for a lowbrow, any of the jokes about the smell of cheese.'

The articles from this period are remarkable for their brevity; whether this was dictated by available space in the *Daily Worker*, or by the time at his disposal I do not know. But in any case, he crammed a remarkable amount of information into a short space.

Haldane followed his own admonition; he always knew much more than is revealed in the article. This is the best way to be able to simplify without unknowingly distorting the scientific truth. An excellent example is 'Breathing', in this volume. Partly this is a textbook account of lung physiology. But it is much more quantitative than most popular articles; for example the surface area of the lung is about 100 square yards. Relying on his own experiments, usually on himself, although he does not say so, he discusses the consequence of breathing too much carbon dioxide. He discusses breathing a 7 per cent concentration. 'If you sit in a chair and breathe as deep and as fast as you can for a minute or so your fingers will probably begin to tingle. If you go on longer you will probably get a cramp in your hands and feet, but some people get convulsions; and probably if every reader of the *Daily Worker* tried it, one or two of them would die.'

An article that I enjoyed dealt with human weakness resulting from our evolutionary past. For instance, there is the nose problem caused by our hypertrophied forebrain: 'Compare a man's nose with that of any other mammal, say a horse, dog, or rabbit. The air going into and out of its lungs has a nearly straight run from the lungs in and out though the nose ... In man each nostril takes a hairpin bend, with very awkward results. When a dog sneezes, the air gets a straight run, and he can clear his nose. A human sneeze cannot get through the narrow and twisted nostrils, so we have to open our mouths.' No wonder we are beset with colds.

Inevitably, some articles now seem dated, though they remain of historical interest. On the subject of computers, he finds it remarkable that a machine (which includes a photo-electric cell) can factor a 21-digit number. No one, not even Haldane, could possibly have foreseen what computers can now do. In an article on cats, he says: 'Tortoiseshell females are common, but males are rare, and we do not know what character their children inherit, or why, as is often the case, they are sterile... If any readers can get me (a tortoiseshell male), I shall be most grateful, and quite prepared to pay. But please write before sending any cats!' Of course Haldane did not know that tortoiseshell males would turn out to be XXY and that the speckled coat pattern in females is caused by differential X-chromosome inactivation in different skin areas.

In the article 'Bird Migration', Haldane confesses to almost total ignorance as to why many birds and Monarch butterflies migrate to specific sites, often at long distances, year after year. He is impressed that young ones find their way without any teaching. Actually, much more is now known about how birds navigate, for example their use of celestial signals. Haldane finishes his article by saying: 'This is one of the problems which is as likely to be solved by ordinary people in their spare time as by laboratory scientists. With the combination of scientific education and leisure to which we may look forward as Leninism spreads over the world, we can look forward to a day when about one person in twenty will be a naturalist, and many mysteries no more.' Well, Haldane could be as much of a romantic as anyone, and as capable of terribly wrong prognostications.

A number of the essays are concerned with the physical sciences. I was struck for the first time by the contrast, which he points out, that while we can hear some half dozen octaves of sound, our range for electromagnetic radiation is barely an octave. Frequencies in the ultraviolet or X-ray range we cannot sense, nor can we see infrared, despite the fact that many animals can. Why can't we? Haldane doesn't know.

One subject on which Haldane saw farther ahead than his contemporaries was DDT. It was confidently assumed that this chemical had no effect on mammals, so all sorts of people, soldiers especially, were dusted with enormous amounts. There were suggestions that this be incorporated into paint, to make a wall permanently fatal to insects. Haldane was suspicious. As was typical, he looked into research showing that DDT was anything but benign. He strongly advocated more testing of compounds before they became widely used, something we now do much more routinely than in Haldane's time. He did not foresee the effects on the food chain, causing the near extinction of some raptors.

Although Haldane assumed that his readers were ignorant of science and went to great pains to keep the articles simple, he did not assume similar ignorance of dialectical materialism. This is probably reasonable, considering the readership of the *Daily Worker.* Many of the articles have a Marxist twist. Haldane was very inventive in finding an excuse for a Marxist sermon. Some are clever, some are contrived, and some are simply annoying to a non-Marxist.

Many of his articles are concerned with activities associated with the war. Haldane, typically, was in the midst of many sorts of activities. He wrote about bomb shelters, explosives, and especially the rescue of sailors from submarines. These experiments, as usual, were done on himself and his associates. The most extensive involved survival at high pressures. Despite Haldane's down-playing the pain and danger, the experiments often were really heroic. One of his students suffered a pneumothorax and Haldane himself suffered a spinal injury that caused him pain the rest of his life. Nevertheless, the Haldane light touch shows up. He claimed to be the only person to have tasted oxygen, which has a taste at sufficiently high pressure. He said it tasted like stale ginger beer.

How well have these articles stood up? Haldane took pride in being up to date in science—all fields. But times change. Most of what he wrote is still correct, though some subjects are no longer the centre of attention. An example of a work that has not gone out of date is his article on cousin marriage. He notes the deleterious effects of in-breeding in general and the enormous increase of rare recessive diseases from consanguineous marriages. Yet, he sensibly points out that the overall excess risk of a cousin marriage is small enough that the decision should be left up to the couple concerned. In contrast, much that he wrote about genetics is no longer in the mainstream of the subject. Much of it is still correct, but a student who comes to genetics via biochemistry is not satisfied with anything not expressed in molecular terms. Yet I would add that many of the problems that Haldane noted as being unsolved are still of interest.

The greatest weakness of Haldane's articles during this period is their subservience to Russian genetics. His Marxism led him to misjudge Stalin and the Soviet regime. In discussing the great scientist Vavilov and his institute, Haldane says: 'The work of this institute was cut down to some extent in the years before the war, largely because the best varieties had been selected, and partly because Lysenko's invention of vernalization rendered many of them less valuable than they were before.' In the first place, Lysenko did not invent vernalization. In the second it certainly did not render a number of varieties less valuable. We now know a great deal more about vernalization and it is indeed a remarkable phenomenon, but Lysenko's propagandizing was nothing short of disastrous. In another article, in a sad example of equivocation, Haldane pointed out that Professor Jeffries of Harvard had attacked Mendelian genetics as forcefully as Lysenko. The comparison is way off the mark. What Haldane did not say was that Jeffries was a harmless eccentric. His opponents were not deprived of their jobs or executed. Incidentally, Vavilov later died in prison.

It is fair to say that, although Haldane adhered to his support of Lysenko much longer than most geneticists, he finally gave up. He stopped writing for the *Daily Worker* in 1950 and, although he retained his Marxist views, he ceased his work for the Party and his support of Soviet genetics dwindled.

I will go beyond Haldane's Communist period and briefly mention his self-obituary. While Haldane was in his final illness, Philip Daly of the BBC had the idea of inviting him to write his own obituary. Haldane relished the chance, and it was broadcast on the day of his death. It begins: 'I am going to begin with a boast. I believe that I am one of the most influential people living today, though I haven't got a scrap of power. Let me explain. In 1932 I was the first person to estimate the rate of mutation of a human gene; and my estimate was not far out. A great many more have been found to mutate at about the same rate since.' He then went on to quite an extensive review of his numerous scientific accomplishments. Immediately after the interview, Haldane returned to his hospital and put the finishing touches on his poem 'Cancer is a funny thing'. It is terrible poetry, but it is a great read.

Some years later, Philip Daly was in Wisconsin to interview Sewall Wright. He told me that in the first draft of his obituary Haldane had said that he was *the* most influential person living today, but he later thought better. The whole episode is vintage Haldane. Whatever you say about Haldane seems to be an understatement.

The series of articles in this book is a very small sample, less than 10 per cent, of Haldane's popular writing during his Marxist period. But they are representative and give an idea of the range of the entire series. The articles are uneven. Some are brilliant, others show evidence of having been tossed off in a hurry. Most were correct, but sometimes he was wrong, at least from a current view. They show the variety of Haldane's character: broad erudition, near-photographic memory, familiarity with the classics, clarity and simplicity of expression, self-assurance, stubbornness, admiration for dialectical materialism, and blind, dogged support of Lysenko. Haldane was all these, and more. In this book you encounter one of biology's most remarkable characters, possibly *the* most remarkable.

There is an old joke, originally told about another person, but which is more appropriate to Haldane. Question: Who is the most widely read biologist? Answer: Haldane—he only had to read what he wrote to have read more than anyone else.

James F. Crow
University of Wisconsin

INTRODUCTION

J.B.S. Haldane's Popular Essays

by Krishna R. Dronamraju
Foundation for Genetic Research Houston, USA

John Burdon Sanderson Haldane, or JBS, was an intellectual giant who possessed the rare quality of making both important contributions to fundamental science as well as to its popularization. He was one of the founders of population genetics, but Haldane defied labels and simple description.

His early science education was provided by his father, the Oxford physiologist John Scott Haldane; the rest was self-taught. Although Haldane never earned an academic degree in science (having interrupted his studies at Oxford to fight in the First World War), he nevertheless became one of the most influential scientists of the twentieth century. He was a polymath whose intellectual contributions had a significant impact on several disciplines including physiology, biochemistry, biometry, cosmology, statistical methodology, and almost all aspects of genetics. His undergraduate classical education at Oxford University sustained his interest in the humanities during his entire life almost as a relaxation from his scientific endeavours.

Haldane's contribution to the social applications of science must be regarded as an important part of his influence on science and society. Early in the twentieth century, he was one of the leading scientists (some have suggested that Haldane and Einstein were alone) who questioned the ethical basis of scientific applications. Early in his career, in 1923, in a remarkable book, *Daedalus, or Science and the Future*, Haldane predicted the future developments in molecular biology and lamented the lack of an ethical framework to facilitate the social acceptance of genetic intervention. Haldane's ideas were later incorporated by Aldous Huxley (1932) in his fictional work *Brave New World.*

As an individual, Haldane could only be described in superlatives. He was a giant, both physically and intellectually. He was a precocious child and a mathematical prodigy. Possessed of a prodigious memory and knowledge of classics, he was equally at home reciting Homer's *Iliad* or discussing the evolutionary consequences of disease-causing genetic mutations. Following in his father's footsteps, Haldane conducted painful physiological experiments upon himself and other volunteers to test the effects of inhaling various gaseous mixtures and ingesting certain chemicals under various experimental conditions. Some of his results were of great importance to diving as well as space medicine.

In Great Britain, Haldane and Julian Huxley were among the best-known scientific popularizers in the biological sciences during the first half of the twentieth century. Haldane's essays covered a wider range of subjects, including animal behaviour, evolution, genetics, geology, paleontology, physics, cosmology, mathematics, medicine, astronomy, meteorology, theology, politics, philosophy, and history. They were published in such varied periodicals as *The Listener,* the *St. Louis Post-Dispatch,* and the *Saturday Evening Post.* During his Marxist years, which spanned the years 1939–50, Haldane's popular essays mostly appeared in the *Daily Worker.* Several years later, he contributed similar articles to newspapers in India. This book includes essays from both periods.

Haldane's writing style

Sir Arthur C. Clarke, well known for his numerous books on science and science-fiction, commented that Haldane was perhaps the most brilliant scientific popularizer of his generation and that 'most of Haldane's scores of essays... may still be read with great profit.'

Haldane's writing style was particularly noted for its simplicity and lucidity. His great skill was evident in his capacity to describe scientific phenomena to the lay public. And he was a prolific writer. During a career spanning over fifty years, Haldane wrote several hundred popular essays, 24 books and more than 400 scientific papers for most of which he was the sole author. Many of his popular essays were written during his travels, usually on the commuter trains and later on planes. He wrote very quickly in long hand, often requiring no corrections or further drafts. It is typical of Haldane's essays that they provide brief glimpses to the reader of a great number of subjects, tantalizing glimpses which are meant to pique one's curiosity but do not provide extensive information. The reader is expected to show initiative to seek fuller information elsewhere.

These essays contain passionate commentaries on a number of scientific and social problems. Their value has not diminished with time. Ethical considerations are discussed wherever relevant. The essays reflect Haldane's itinerant optimism that scientific applications are generally beneficial to society. They provide us with an excellent opportunity to evaluate how Haldane saw the world and its problems and to what extent science can provide solutions.

In an essay entitled 'Common sense about the planets', Haldane emphasized that in some instances, as in astronomy, while data about the solar system continued to be collected over centuries, a rational explanation was only possible when technology advanced sufficiently. His point was that it is futile to reach conclusions on the basis of incomplete data. It is preferable to say that we do not know the explanation.

On the other hand, in an essay on bee communication, Haldane described in detail what was known about social behaviour among honey bees. Of special interest is the meaning of bee 'dances' in communicating the direction and distance to a food source. He wrote: 'These observations seem to have a great philosophical importance. It is often said that animal "language" is a mere expression of the emotions, and cannot convey statements of fact. But it is clear that the bees can tell each other not merely that they have found food, but where they have found it.'

These two subjects, as well as numerous others, were selected by Haldane to illustrate diverse situations in science.

Ethical issues receive serious attention in several essays, for instance, in 'Some reflections on non-violence', 'The sound of one hand', and 'Darwin in Indian perspective'. Writing on the impact of Darwinism, Haldane wrote that Darwin was led to the theory of natural selection by reading Malthus. Consequently, although Darwin emphasized the role of physiological adaptation, he left the erroneous impression that the struggle for life and survival of the fittest were analogous to war and economic competition in human societies. And Darwinism was misused to justify such activities.

Haldane's essays can be enjoyed by all scientists and educated laymen alike. They should be of special interest to historians of science as well as those interested in teaching courses in science education and science writing.

Acknowledgements

I am indebted to N. Avrion Mitchison and Lois Godfrey, Professor Haldane's nephew and niece respectively, for permission to reproduce these essays. I am most grateful to Sir Arthur C. Clarke and Professor James F. Crow for providing the Foreword and Preface respectively. I am most appreciative of the assistance provided by Latha Menon and Jim Thompson at Oxford University Press while preparing the manuscript. Michele Wambaugh assisted in the photographic work.

I am grateful to the *Daily Worker* (now *Morningstar*) for permission to reproduce the essays from Haldane's Marxist years (1939–50), and *The Hindu* for permission to reproduce the essays written during his life in India (1957–64).

FURTHER READING

By J. B. S. Haldane

1923 *Daedalus, or Science and the Future* (London: Kegan Paul, Trench, Trubner).

1929 'The origin of life', *Rationalist Annual*, p. 3.

1932 *The Inequality of Man and Other Essays* (London: Chatto & Windus).

1932 *The Causes of Evolution* (London: Longmans, Green).

1934 (with J. R. Baker) *Biology in Everyday Life* (London: George Allen & Unwin).

1935 *Science and Well-being* (London: Kegan Paul, Trench, Trubner)

1938 *Heredity and Politics* (London: George Allen & Unwin).

1940 *Science in Everyday Life* (New York: Macmillan).

1940 *Keeping Cool and Other Essays* (London: Chatto & Windus).

1941 *New Paths in Genetics* (London: George Allen & Unwin).

1947 *Science Advances* (London: Allen & Unwin).

1947 *What is Life?* (New York: Boni & Gaer).

1951 *Everything Has a History* (London: Allen & Unwin).

1961 'An autobiography in brief', *Illustrated Weekly of India*, Bombay (reprinted in *Selected Genetic Papers of J. B. S. Haldane*, ed. K. R. Dronamraju (New York: Garland, 1990).

1964 'A defense of beanbag genetics', in *Perspectives in Biology and Medicine*, 7: 343–59.

1968 *Science and Life: Essays of a Rationalist*, introd. J. Maynard Smith (London: Pemberton Publishing).

By others

Crow, J. F. (2004) 'Haldane's ideas in biology with special reference to disease and evolution', in K. R. Dronamraju, ed., *Infectious Disease and Host-Pathogen Evolution* (Cambridge: Cambridge University Press), pp. 11–17.

Clark, R. W. (1968) *JBS: The Life and Work of JBS Haldane* (New York: Coward-McCann).

Dronamraju, K.R., ed. (1968) *Haldane and Modern Biology* (Baltimore, MD: Johns Hopkins University Press).

Dronamraju, K.R., ed. (1985) *Haldane: The Life and Work of JBS Haldane with Special Reference to India* (Aberdeen: Aberdeen University Press).

——, ed. (1995) *Haldane's Daedalus* Revisited (Oxford: Oxford University Press).

—— (2006) 'J.B.S. Haldane (1892–1964)', in K. R. Dronamraju and P. Arese, eds, *Malaria: Genetic and Evolutionary Aspects* (New York: Springer), pp. 13–24.

AN AUTOBIOGRAPHY IN BRIEF

J. B. S. HALDANE

A number of inaccurate statements have been published about me in the press—quite as many in articles favouring me as in hostile statements. So I propose to give a brief account of my scientific career. I was born in 1892. I owe my success very largely to my father, J. S. Haldane. He was perhaps best known as a physiologist, but he was so far from being a specialist that in later life he was elected president of the Institution of Mining Engineers and delivered the Gifford lectures on the 'Existence and Attributes of God'. I suppose my scientific career began at the age of about two, when I used to play on the floor of his laboratory and watch him playing a complicated game called 'experiments'—the rules I did not understand, but he clearly enjoyed it.

At the age of eight or so I was allowed to take down numbers which I called out when reading the burette of a gas-analysis apparatus and later to calculate from these numbers the amounts of various gases in a sample. After this I was promoted to making up simple mixtures for his use and, still later, to cleaning apparatus. Before I was 14, he had taken me down a number of mines, and I had spent some time under water both in a submarine and in a diving dress. He had also used me as the subject in many experiments. In fact I spent a good deal of my holidays from school in learning my father's trade. Most Indian boys do this, but not the sons of scientists. After I was 12, he discussed with me all his research before publication, and sometimes tried out a lecture course on me before delivering it to students.

At school I deserted 'classics', that is to say, the study of Latin and Greek, at the age of 14 and studied chemistry, physics, history, and biology, with my father's full backing but to the annoyance of the headmaster, who said I was becoming 'a mere smatterer'. The teaching of chemistry was good, and by the age of 16 I had learned some facts discovered since my father had studied that subject, so that I could help him and C. G. Douglas; and my first scientific paper was a joint one with them, read to the Physiological Society when I was 17.

I went to Oxford on a mathematical scholarship in 1911 and took first-class honours in mathematical moderations (roughly the Indian BSc level). But as nobody can study mathematics intensively for more than about five hours daily and retain sanity, I also attended the final honours course in zoology in my first year. One of my fellow students was the late Professor Narayan K. Bahl, who later did so much for the teaching of zoology in India. At a seminar for zoölogy students in 1911, I announced the discovery, from data published by others, of the first case of what is now called linkage between genes in vertebrates. My evidence was considered inadequate, and I began breeding mice with A. D. Sprunt, who was killed in 1915. In 1912 I switched over to *literate humaniores*, a course based on Latin and Greek classics, but including the study of a good deal of modern philosophy and ancient history. I took first-class honours in this subject in 1914 and had intended to go on to study physiology. But in 1914 I joined the British army and have, therefore, no scientific degree. In 1915 my mouse work with Sprunt and my sister, Mrs Mitchison, was published.

During the First World War, I was wounded twice, in France and in Iraq, after which I spent 16 months in India. I determined to come back as soon as I could associate with Indians on a footing of equality.

On returning to Oxford after the war, I was elected a Fellow of New College and began teaching physiology while myself attending Sherrington's advanced practical course in that science. Indian readers who find it incredible that I was appointed without a degree in physiology, or any other science, would do well to remember that Srinivasa Ramanujan, India's greatest mathematician since Aryabhatta, had no degree and would thus be disqualified from teaching in an Indian University were he alive today. I may not have been a good teacher, but I was a successful one. In 1922 there were about 60 candidates for honours in physiology, three from New College. These three were one-half of the six who secured first-class honours. I had 20–30 hours a week of teaching and other university duties. However, I managed to get ten hours of private tuition done at night after 8 p.m. and to concentrate 9 hours on Wednesday, so I got some time for research and reading.

I worked on human chemical physiology and on genetics. Perhaps my most important discovery in physiology was that when I drank ammonium chloride solution I developed various symptoms of severe acid poisoning, including breathlessness. My main genetic discovery was the rule as to the sex of hybrid animals. In 1921 I put in a term as biochemist in the Edinburgh Royal Infirmary and learned a little medicine in the wards.

In 1922 Professor Hopkins (later president of the Royal Society, Nobel laureate, etc.) invited me to Cambridge as reader in biochemistry. I was his second-in-command for 10 years and supervised the work of about twenty graduate students—much of which was first rate. Perhaps my own most important discovery was that a substance for which carbon monoxide competes with oxygen, now called cytochrome oxidase, was found in plant seedlings, moths, and rats. The most remarkable thing about this discovery was that I was able to find out a good deal about a substance in the brains of moths without cutting them up or killing them. However, my enunciation of some of the general laws of enzyme chemistry may have been more important.

In 1924 I published what my colleagues generally think my most important paper, the first of a series on the mathematical theory of natural and artificial selection. Five of these papers have been reprinted in the United States and are available to libraries which do not possess the *Proceedings of the Cambridge Philosophical Society.*

They contained calculations showing great intensity of natural selection in favour of dark colour in a British moth species. This was regarded as ridiculously high, but 30 years later Kettlewell found a slightly higher figure in field studies. In 1930, in my book *The Causes of Evolution*, I published the first estimate of a human-mutation rate. Since then, this has become a matter of international politics in connection with atom-bomb tests.

Towards the end of my period at Cambridge I spent some time at the John Innes Horticultural Institution in a London suburb, directing research on plant breeding, and continued to do so after I became parttime professor of genetics in University College, London, in 1933. My most important work was with Miss de Winton on an ornamental plant, *Primula sinensis.* We were the first, for example, to study linkage in a plant with double the usual number of chromosomes. I also showed that one of the genes responsible for its colour acted by changing the acidity of the petal sap.

I have always been of some use to my colleagues because I knew what was going on in several different branches of science, and it was, I think, in the autumn of 1933 that I did what posterity may regard as the best and most important action of my life. I found posts for several Jewish refugees from Germany, and I did my best to help others. One evening Dr Boris Chain dined in my house. We talked about the work he had done in Germany, and I said, 'There is a man named Florey at Oxford who is interested in that sort of thing. I advise you to visit him.' Later Florey and Chain isolated penicillin, which has saved hundreds of times more lives than atomic bombs have caused deaths, a fact often forgotten by critics of science. Florey and Chain have been rewarded for this work. They shared a Nobel prize, and Florey is now president of the Royal Society. Perhaps all my discoveries will be forgotten and I shall be remembered only in the words of the ancient Greek poet Pindar: 'He once nourished the contriver of painlessness, the gentle limb-guardian Asklepios (Dhanvantari), the heroic conqueror of manifold diseases.' Bacteriologists, by the way, are heroic: bacteria are much more dangerous than tigers.

For such activities I had the honour of figuring on the list of persons to be arrested if German armies conquered England in 1940.

In 1936 I became professor of biometry in London but never got a building for my own use. Some of my colleagues in this department did very fine work, and two became Fellows of the Royal Society, as I had in 1932. I participated in a little of their work and made some contributions to mathematical statistics, of which perhaps the most labour-saving is my calculation of the cumulants of the binomial distribution.

In 1939 the British submarine '*Thetis*' sank on her trials with the loss of over one hundred lives. About one-half the dead were civilians; two unions asked me to investigate the disaster. I did some experiments on myself and friends, no more drastic than I had done at Cambridge, and shed enough light on what had happened to convince the British Admiralty that their 'experts' knew very little. They asked me to continue the experiments, and when war broke out I was given various assignments. E. M. Case and I, for example, were the first people to pass 48 hours shut up in a miniature submarine with apparatus which we had correctly calculated would renew the air for that time. My wife and I worked out methods for the rapid ascent of divers, and so on. During this work I made a curious discovery. Oxygen, when breathed at a pressure over about 6 atm., has quite a taste. Nevertheless, since text-books have priority over truth, students of chemistry are well advised, when examined, to state that 'oxygen is a colourless, inodorous, and tasteless gas.' I advise even M.Sc. students against stating that Case and Haldane reported to the contrary in a letter to *Nature* in 1941. After breathing oxygen for five minutes or so at such pressures, one has violent convulsions; and my frequent demand for a soft chair or a cushion is due to the fact that I fractured my backbone in such convulsions.

Among the papers I wrote during the 12 years between the end of the war and my departure to India is one published in 1956 on a method for estimating the number of lethal mutations produced in mice by gamma rays and other agents causing mutation. A clear answer to this question would allow us to give a partial answer to the question. 'How many human babies in future generations will die as the result of atomic-bomb tests?' I was not, of course, offered facilities for such work. G. S. Carter at Harwell began to use my method. He then resigned his post, for an undisclosed reason, and took a job with a poultry-breeding firm. However, Sugahara, Okazawa, Tutikawa, and Muramatsu have used heavy doses, like those absorbed by the survivors of Hiroshima and Nagasaki. The Japanese workers have used much smaller doses, such as might be given to workers in atomic-energy establishments or to radiologists who took precautions. They naturally got rather few mutations and cannot yet estimate the rate very accurately. According to my method, about 500 rs are needed to produce a lethal mutation, while two other methods give a somewhat lower figure of about 300, which was what I guessed in 1956. If the Japanese workers are right, the damage done to future generations by the tests so far carried out is a bit less than Pauling and Russell have stated but very much more than American official spokesmen have claimed.

In 1957 I came to India to work at the Indian Statistical Institute, and I have to thank Professor P. C. Mahalanobis for making this possible. My most important work there was, beyond doubt, starting S. K. Roy, K. R. Dronamraju, T. A. Davis, and S. D. Jayakar on their scientific careers, which are likely, in my opinion, to be illustrious. At least 20 of my pupils have become Fellows of the Royal Society, so I can probably judge fairly well. At the Indian Statistical Institute I personally published two pieces of theoretical work which may be of lasting value, besides many which are unlikely to be so. Since leaving it, and while employed by the Council of Scientific and Industrial Research, I have published jointly with Jayakar one paper on human relationships. I am grateful to Mahalanobis for giving me the opportunity of working in the Indian Statistical Institute, where I learned a great deal about what can and cannot be done in India, even though it gradually became clear that I could not carry out the kind of work I had wanted to in that institute.

I have of course done a good deal more than appears in this summary. I have taken part in politics, written a book of stories for children, and put my Latin and Greek learning to some use by commenting on biological passages in ancient writings. I may have been the first to ask the cosmological question, 'Is space-time simply connected?'—though only a man of the stature of Einstein is likely to answer it. And I have made several bad mistakes. But I think this article gives some notion of my contributions to scientific knowledge.

PART I

Essays from the Marxist period (1937–50)

I

How to write a popular scientific article

Most scientific workers desire to spread a knowledge of their subject and to increase their own incomes. Both can be done by writing on science for the general public. In what follows I shall give some hints on how to do it. But let no reader suppose that my method is the only one. Literary synthesis is like organic chemical synthesis. The method to be adopted depends on the product required, the raw materials, and the apparatus available. As my brain is my apparatus, and different from yours, my methods will also differ from yours.

The first thing to remember is that your task is not easy, and will be impossible if you despise technique. For literature has its technique, like science, and unless you set yourself a fairly high standard you will get nowhere. So don't expect to succeed at your first, or even your second, attempt.

For whom are you writing? This is even more important than the choice of subject. For you will not get an article on the history of eighteenth-century physics into a daily newspaper. Moreover the length of your article will depend on where it is to be published.

Now for the subject matter. You may take a particular piece of research work, or a particular application of science. Or you may choose some general principle, and illustrate it from different branches of scientific work. For example an excellent article could be written on fruitful accidents. Priestley broke a thermometer, and the fate of the mercury from it led him to the discovery of oxygen. Takamine spilled some ammonia into a preparation from suprarenal glands, and crystallized out adrenaline. Probably you will do better to begin on some more specialized topic, unless you are a student of the history of science.

Remember that your treatment of it must be highly selective. So far you have probably written two main types of article. Firstly, answers to examination questions in which you tried to show how much you knew about some topic. And secondly, scientific papers or technical reports which dealt very exhaustively with a small point. Now you have to do something quite different. You are not trying to show off; nor are you aiming at such accuracy that your readers will be able to carry out some operation. You want to interest or even excite them, but not to give them complete information.

You must therefore know a very great deal more about your subject than you put on paper. Out of this you must choose the items which will make a coherent story. A number of the articles which are submitted to me from time to time are far too like examination answers. They give the impression that the author has looked his subject up, and tried to give a condensed summary of it. Such a summary may be all very well in a text-book, but will not hold the attention of a reader of popular articles, who does not contemplate severe intellectual exertion.

This does not mean that you must write for an audience of fools. It means that you must constantly be returning from the unfamiliar facts of science to the familiar facts of everyday experience. It is good to start from a known fact, say a bomb explosion, a bird's song, or a cheese. This will enable you to illustrate some scientific principle. But here again take a familiar analogy. Compare the production of hot gas in the bomb to that of steam in a kettle, the changes which occur in the bird each year to those which take place in men once in a lifetime at puberty, the precipitation of casein by calcium salts to the formation of soap suds. If you know enough, you will be able to proceed to your goal in a series of hops rather than a single long jump.

If you try to write an article in this way, you will probably discover your own ignorance, especially of quantitative matters. How completely do a robin's gonads revert to an infantile condition in autumn? How much more calcium is there in milk than in London tap water? What is the maximum temperature in an exploding bomb? It may take you twelve hours' reading to produce an intellectually honest article of a thousand words. In fact you will have to educate yourself as well as your public.

When you have done your article, give it to a friend, if possible a fairly ignorant one. Or put it away for six months and see if you still understand it yourself. You will probably find that some of the sentences which seemed simple when you wrote them, now appear very involved. Here are some hints on combing them out. (Remember, by the way, that I am only giving my personal opinions. Professor Hogben writes sentences longer than some of my paragraphs, and his books sell very well, as they ought to.) Can you get in a full stop instead of a comma or a semicolon? If so, get it in. It gives your reader a chance to draw his breath. Can you use an active verb instead of a passive verb or a verbal noun? If so, use it. Instead of 'It is often thought that open windows are good for health', or 'There is a widespread opinion that open windows are good for health', try 'Many people think that open windows are good for health.' Or 'Most people', if you think that is the case.

Try to make the order of the phrases in your sentence correspond with the temporal or causal order of the facts with which you deal. Instead of 'Species change because of the survival of the fittest' try 'The fittest members survive in each generation, and so a species changes.' Not that I like the phrase 'a species changes'. It would be better to say 'the average characters of the members of a species, such as weight or hair-length, change'. Of course in the history of scientific discovery an effect is commonly known before its cause. And fairly often a mathematical theorem is known to be probably true before it is formally proved. If you enunciate your theorem before you prove it you are apt to give the impression, as Euclid does, that you are producing rabbits from a hat. Whereas if you lead up to it gently you create less impression of cleverness, but your reader may find your argument much easier to follow.

In a scientific, and still more, a mathematical paper, elegance of presentation, which often means the hat-and-rabbit method, is always great fun, and sometimes desirable. How delightful to produce some wholly unexpected function at the last moment by contour integration, to damn a suggested mechanism by an appeal to Hearnshaw's theorem, or to label a plant which won't breed true as just another case of balanced lethals. By doing so you may help the serious student to short cuts in thinking. But you will merely dazzle the ordinary reader. Go slow, and show him as many steps as you can in your arguments or causal chain, even if, in your own thinking, you skip some of them or take them backwards.

When you have written your article it may seem rather gaunt and forbidding, a catalogue of hard facts and abstract arguments. A critic may say it needs padding. I object to padding for padding's sake. It is characteristic of writers who are more interested in their style than their subject matter, such as Charles Lamb or Robert Lynd, but out of place in a scientific article. On the other hand you must do what you can to help your reader to link up your article with the rest of his knowledge. You can do this by referring to familiar facts or to familiar literature. I have been severely criticized for 'dragging in' references to Marx in my articles in the *Daily Worker*, though I think I refer to Engels more frequently. But a number of my readers are familiar with the works of these authors. Engels said certain things about change, as Heraclitus said very similar things before him, and Bergson and Whitehead after him. But for one of my readers who has read Heraclitus, Bergson, or Whitehead, a hundred have read Engels, so I prefer to cite him. If I were lecturing on the same matters to classical scholars I should perhaps site Heraclitus, even though I think Engels said it better.

In my last book on genetics, there are seven quotations from Dante's *Divine Comedy*. I have been criticized for 'dragging in'

Dante. But I think it worth while to show the continuity of human thought. I don't agree with Dante's theory that mutations are due to divine providence, but I consider it desirable to point out that he had a theory on this subject. I think that popular science can be of real value by emphasizing the unity of human knowledge and endeavour, at their best. This fact is hardly stressed at all in the ordinary teaching of science, and good popular science should correct this fault, both by showing how science is created by technology and creates it, and by showing the relation between scientific and other forms of thought.

A popular scientific article should, where possible, include some news. I try, as a general rule, to include one or two facts which will not be familiar to a student taking a university honours course in the subject in question, unless his teachers keep well up with the periodical literature. As there is often a lag of five years between the publication of a discovery and its inclusion in a textbook, this is not very difficult in peace time. But it is not very easy at present, in view of the number of libraries which have closed down, and the absence of many European and some American peridicals. Of course some care is needed in appraising new work. A very large number of alleged discoveries are not confirmed by subsequent workers. One well-known English popularizer of science has a perfect genius for picking out discoveries of this kind for announcement to the public. If, like myself, the writer is actually engaged in research, and has seen a number of his own bright ideas go west, he is less likely to fall into this particular trap.

In the early stages of popular writing it is well to write out a summary of the article, though I rarely do so myself. Here is a possible skeleton for an article on cheese.

Introduction. A well-known fact, say the shortage of cheese.

Central theme. The process of cheese manufacture.

Why it is important. Cheese as the cheapest food containing large amounts of 'good' protein. Vitamins and calcium in cheese.

Connections with other branches of science. Rennet compared with other enzyme preparations used in industry, e.g. in confectionery and tanning. Other uses of specific microorganisms, e.g. in brewing. Why putrid cheese is safer than putrid meat.

Practical suggestions. How to increase our cheese output. Combating mastitis in cows. Cattle feed and fertilizers. Should cargo space be devoted to cheese rather than meat? Need for scientific planning of national food supply.

How much of this you can get in depends on the length of your article and your capacity for compession. If you are writing for a highbrow journal you may quote the passages on cheese from Euripides' 'Cyclops', if for a lowbrow, any of the jokes about the smell of cheese.

That is one way of doing it. But other writers would show cheese as part of the Mysterious Universe. We do not understand protein synthesis, not the extreme specificity of some enzyme actions. Cheese-making is part of the pre-scientific activities by which we still keep a communion with nature. Cheese is a natural food, and beef is not. And so on. I think this is an anti-scientific attitude. But you can sell that sort of stuff, and get over a certain amount of genuine knowledge while doing so. Everyone must write popular scientific articles in his own way. I have only described one way, and I do not claim that it is the only way, or even the best possible way.

2

Why I am a materialist

When I say that I am a materialist I mean that I believe in the following statements:

1. Events occur which are not perceived by any mind.
2. There were unperceived events before there were any minds.

And I also believe, though this is not a necessary logical deduction from the former two, that:

3. When a man has died he is dead.

Further, I think that it is desirable that other people should believe these statements. I do not mean that I believe that the universe is a machine, nor that I am a machine; nor yet that consciousness does not exist, or has a lesser reality (whatever that means) than matter. When I say 'I believe' I do not mean the word in the sense in which a fervent Christian uses it concerning the Virgin Mary, Pontius Pilate, and others who figure in the creeds. I mean it in the ordinary sense, in which, for example, I believe that dinner will be waiting when I go home, though of course the cook may go on strike or the chimney may catch fire. That is to say I act, and propose to act, on the basis that materialism is true. But I am prepared to consider evidence to the contrary. And I certainly don't get shocked or angry if someone criticizes or doubts the truth of materialism.

Now the word 'materialism' is used, particularly in controversy, to imply a belief that a good dinner is better than a good deed. In fact, a materialist is supposed to be a man who has, or does his best to have, large meals, a large mistress, a large bank balance, a large motor-car, and so on. It is not obvious why this should be so. Other people's meals are as material as mine, and a bank balance is not something tangible, like a cellar full of gold and jewels.

In practice I have found that professed materialists are generally less selfish than professed idealists. For idealism is a remarkably useful device to enable us to bear other people's ills, and particularly their poverty. It is easy to persuade ourselves that the poor have various spiritual blessings. But it is not so easy, when one's own affairs are concerned, to avoid the attitude of the idealist of whom it was written:

> There was a faith-healer of Deal
> Who said: 'Although pain isn't real,
> When I sit on a pin and it punctures my skin
> I dislike what I fancy I feel.'

I do not of course deny that some idealists are excellent people, and some materialists coarse and selfish. But on the whole I think the contrary is true, for reasons which will appear later.

Fifteen years ago I was a materialist in practice, but not in theory. I treated myself as a material system. We all do this to some extent. When we want to go somewhere we get into a train or bus, confident that on the one hand we shall not be able to propel ourselves so rapidly through space by the mere exercise of our wills, nor on the other that the vehicle will find any more difficulty in moving us than if we were a sack of potatoes. However, though we all have considerable faith in the applicability to ourselves of the laws of physics, our faith does not apply to chemistry. We should be willing to trust our weight to a rope which had been tested to stand double our weight; but we should mostly hesitate to drink half the fatal dose of a poison. Rightly too in some cases, for poisons in sub-lethal doses may do a good deal of harm. But not by any means always. Some poisons, such as carbon monoxide, are completely harmless in half the lethal quantity.

I applied the laws of chemistry to myself. For example, I said: 'If a dog is given hydrochloric acid to drink (diluted of course so as not to injure its stomach) it excretes part of the acid combined with ammonia as ammonium chloride. Now men work in a similar way to dogs, and both are systems of partially reversible chemical reactions. So if I eat ammonium chloride I shall become more acid.' This did in fact happen. I was quite correct in my reasoning, or at any rate it led to a correct result.

However, although I was a materialist in the laboratory, I was a rather vague sort of idealist outside, for the following reason. I had learned that matter had certain properties. It consisted of atoms which united in particular patterns. They moved in definite paths under given forces, and so on. My belief in these theories was not a matter of mere docility either. I had tested them and risked my life on their substantial accuracy. Clearly, if matter had the properties attributed to it by physicists and chemists, something more was needed to account for living organisms. And it was far harder to account for mind. As a believer in evolution I had to reject such theories as T. H. Huxley's epi-phenomenalism, according to which mind is a secondary consequence of a small class of material events (namely, those which go on inside our heads), but does not influence them. Apart from my very strong belief that I can act, the evolution of something as complicated as my mind, yet absolutely functionless, seemed most unlikely. Not that functionless organs are never evolved. On the contrary, it is probable that most organs are evolved in a rudimentary form before they develop a function. And I have not enough faith in the theories of Paley and his like to believe that every organ—for example, a cock's comb, a pigeon's cere, or a cassowary's wattle—has a function. However, I cannot believe that a system so complicated, and within its limitations so efficient, as the human mind could have evolved if it were functionless.

Nor did I see how, on a materialist basis, knowledge or thought was possible. The light which reaches my eyes causes nervous impulses in about half a million fibres running to my brain, and there gives rise to sensation. But how can the sensation be anything like a reality composed of atoms? And even if it is so, what guarantee have I that my thoughts are logical? They depend on physical and chemical processes going on in my brain, and doubtless obey physical and chemical laws, if materialism is true. But if so I have no reason for thinking that it is true. So I was compelled, rather reluctantly, to fall back on some kind of idealistic explanation, according to which mind (or something like mind) was prior to matter, and what we call matter was really of the nature of mind, or at least of sensation. I was, however, too painfully conscious of the weakness in every idealistic philosophy to embrace any of them, and I was quite aware that in practice I often acted as a materialist.

The books which solved my difficulties were Frederick Engels' *Feuerbach* and *Anti-Dühring*, and later on V. I. Lenin's *Materialism and Empirio-criticism*. But the actual progress of scientific research in the last fifteen years also helped me enormously. None of the books which I have mentioned is easy if one has been brought up in the academic tradition which goes back to Plato and Aristotle. This is partly because they apply scientific method not merely to philosophy, but to philosophers. They are not only concerned with showing that their authors are right and their opponents wrong, but with explaining why, under particular social conditions, such and such theories are likely to gain wide acceptance. Hence, unless one accepts their political and economic theory, one is not likely to agree with their views concerning nature and knowledge, though it is only with the latter that I am concerned in these pages.

Engels and Lenin were firm materialists—that is to say, they believed that matter existed before mind, and that our minds reflect nature, and reflect it truly up to a point. But they absolutely rejected the current scientific theories of their day as complete or even satisfactory accounts of nature.

> The sole property of matter [wrote Lenin], with whose recognition materialism is vitally connected, is the property of being objective reality, of existing outside of our cognition.... The recognition of immutable elements, the immutable substance of things, is not materialism, but metaphysical, anti-dialectical materialism.... It is of course totally absurd that materialism should... adhere to a mechanistic world-picture of matter and not an electro-magnetic or some immeasurably more complicated one.

Writing of the physics of his own day, he said: 'Dialectical materialism insists on the temporary, relative, approximate character of all these milestones on the road of knowledge of nature.'

Nature is in a state of perpetual flux—in fact, it consists of processes, not things. Even an electron is inexhaustible—that is to say, we can never give a complete description of it. We professors are always trying to give such a complete description, so that we can deduce all natural happenings from a few general principles. These attempts are successful up to a point, but we always find that nature is richer than we had thought. And the newly discovered properties of things appear to us as contradictions. Thus at the present moment both light and matter are found to have two sets of properties—one set resembling those of particles, and another set resembling those of waves. According to Engels and Lenin, things really embody a union of opposites, whose struggle makes them unstable and results in their development into something else. When we find 'internal contradictions' in our conceptions about things our minds are mirroring nature.

But these internal contradictions do not mean that nature is irrational. They mean that it is unstable. Our brains are finite. Nature is probably infinite, certainly too large for us to take in. So our account of any material phenomenon is a simplification. We naturally think of things as neatly rounded off, and therefore tend to exaggerate their stability. However, the more we study nature the more we find that what is apparently stable turns out to be the battlefield of opposing tendencies. The continents are the field of a struggle between erosion, which tends to flatten them, and folding and vulcanizing, which build mountains. For this reason they have a history. Animals and plants are never completely adapted to their environment, as Paley thought, and as they presumably would have been had they been made by an all-wise and all-powerful creator. On the contrary, they evolve just because they are imperfect. The same principle holds for human societies.

One of the materialist's greatest difficulties used to be perception. If the world consists of self-contained objects isolated from one another in space, how can any sort of image of it be formed in our brains? There is no hollow space in our heads where a puppet representation of the external world could be set up. Sound is the only feature of the external world about whose representation in our brain we know much. If we place an electrode on the auditory part of a cat's cerebral cortex and another somewhere else on its body, then in favourable circumstances if we amplify the current between them and pass it through a loud-speaker we actually hear sounds which the cat is hearing, or would hear were it fully conscious. The same experiment is quite possible with a conscious human brain, though I don't think it has yet been done.

This means that the ear and the auditory nerve serve to set up electrical disturbances in one part of the brain with the same periods as the disturbances in air which we perceive as sound. In this case, then, there is an actual imaging of the external reality. But how can anything of this kind take place with a solid object seen or felt? The physical discoveries of the last decade have shown that ordinary material objects, from electrons upwards, can be regarded as periodic disturbances. Certainly the rhythm is very much faster than that of sound, and could not possibly be copied in the brain. But some kind of rhythmical changes in the brain, though very much slower than those which they mirror, would be copies of at least one aspect of matter.

The physicists tell us that the frequency of the vibrations associated with a particle are proportional to its mass, and the physiologists, in studying the impulses in a nerve fibre from an end organ responsible for our touch or pressure sense, find that the frequency of the impulses increases with the stimulus, though not in exact proportion. We do not yet know in any detail what happens in the brain when we feel pressure, but it is likely that a similar law holds good.

We are only on the very fringe of the necessary investigations, but it is becoming daily more plausible that our minds are physical realities acted on by the rest of the world and reacting on it. Our minds are processes which occur in our brains. Until recently it was quite impossible to see how the processes going on in thousands of millions of cells could possibly form a unity such as we find in our consciousnesses. We are now, however, discovering both in atoms and molecules properties of a system as a whole which cannot be located at any particular place in it. There is nothing in any way mystical about these properties. They can be very precisely measured and calculated. They are expressions of the fact that the various constituents of nature are much less isolated than was at one time thought.

The difficulties about truth are complicated by the fact that we use the word for at least three very different relations. We may mean that a perception or idea in a mind is true if it corresponds to an external reality. If the relation between the two is one of likeness it can never be complete, but it may be true enough for a particular purpose. We may mean that a physical copy or image is like its original. Or we may mean that a statement is true. This statement may be in words or other symbols, and logic is largely concerned with the truth of statements. Their truth or otherwise depends on the meaning of the symbols. This is a social matter. A statement is true only as long as someone understands it. After that it is meaningless. 'Iron is heavier than water' will be true as long as someone understands English, even if he is only an antiquarian. After that it will be gibberish like 'Pung twot maboruulioo', which for all I know meant something to the men who built Stonehenge, but is neither true nor untrue to-day.

Of course the philosophers say that a symbolic statement stands for a mental reality called a judgement, which is independent of language. I think this is extremely doubtful. On the contrary, it seems much more likely that language began with words or phrases whose English equivalents would be 'Come here', 'Wolf!', 'Heave-ho,' 'Darling!', and so on, which are not statements, and neither true nor false. And one can certainly think without making statements or judgements, as when one remembers the plan of a town and picks out the quickest route, or imagines what an acquaintance will do in given circumstances.

The great advantage of the theory that judgements are anything but sentences repeated in our heads is that it gives philosophers a chance to theorize about thought without investigating the physiology of the brain. This enables them to tell us a lot about truth, but very little about how we get to know it or how we act on it. If we take the view that a statement is true in so far as it calls up mental images which correspond to reality, and useful in so far as it cites actions appropriate to the real situation, we have got away from metaphysics, and are up against problems concerning the action of the brain, the history of language, and how we learn language as children, which cannot be solved by pure thought, but only by studying the real world.

For such reasons as these I find materialism intellectually satisfactory. I also think it is useful because it leads to actions of which I approve. Mankind is up against a very difficult situation. We have dealt with a great mass of problems in the past by scientific thinking—that is to say, materialistic thinking. We try to solve our political problems by appeal to eternal values. But if we start thinking materialistically about these 'eternal values' we find that they are social phenomena which have come into being in the last few thousand years, because men gave up hunting and took to husbandry, agriculture, and handicraft. So society became a great deal more complicated, and 'eternal values' are part of the apparatus by which it has been kept going. In particular they are very useful to those who are in comfortable situations at present, and would like the present state of things, with a few minor modifications, to be eternal.

Materialistic thinking in the past has been revolutionary in its effects. It has built up natural science and undermined religion. The same process is going on today. We have to realize that our current ideas about society are mostly very like our ancestors' ideas about the universe four hundred years ago—irrational traditions which stifle progress in the interests of a small minority. These ideas are being transformed by materialistic thinking about history as our ancestors' ideas were transformed by materialistic thinking about nature. The consequence will no doubt be revolutionary, as it was in the past. This would perhaps be deplorable if our society were working well. But it is working very badly. So we are probably going to have an uncomfortable time in the immediate future, whatever happens. And as I want a rational society to come out of our present troubles I am not only a materialist myself, but I do what I can to make other people materialists.

3

Adventures of words

One of the first things a Marxist learns is that everything has a history. Nothing has existed for ever in its present form. Some things have a long and important future, others are going to perish very soon. Thus the biologist can trace the ancestry of men and bisons, and say, that, as there are 2,000 million men, of whom about one-tenth are already seriously planning their future, and only about two thousand bisons, preserved as curiosities, men have more of a future than bisons.

The sociologist can compare the young and vigorous Socialist state and its institutions, such as soviets, collective farms, and state planning commission, with capitalist states and their institutions such as stock exchanges and hereditary titles, and also with more primitive societies such as African tribes and their polygamous chiefs and rain-making wizards. And he can forecast where the future hope of mankind lies.

Today I am concerned with the history of words. Some merely change their form, but not their meaning. The stablest of all are numbers. 'Four' and 'seven' have changed a little from their Anglo-Saxon forms 'feower' and 'seofon', but their meaning has not changed.

One might think that words for common and easily recognized things, such as animals, could not change their meaning. But some are doing so today. A hundred years ago a kid generally meant a young goat. Now it generally means a human child. Quite likely the old meaning will be forgotten in a hundred years. The Anglo-Saxon word for cattle was 'feoh'. As payments were commonly made in cattle it also came to mean a payment, and has survived in the modern 'fee'. It is also the root of such words as fief, feudal, and feu, which is the Scottish word for a lease. They all refer to payments in kind by peasants who never saw a coin in their lives. On the other hand our word 'cattle' is derived from the Latin 'capitale', meaning stock, from which the word 'capital' also comes. If our language had developed a little differently, Marx's greatest work would be called 'Cattle'.

Naturally names describing social position change as society changes. In an Anglo-Saxon house the husband was called 'hlaf-weard', or loaf-warden, his wife the 'hlaefdige' or loafkneader, and the servant 'hlafaeta', or loaf-eater. 'Hlaf' is derived from the same root as the Russian 'khleb', for bread, and 'dige' from the same as 'dough'. The first of these words have gone up in the world, and become 'lord' and 'lady'. Perhaps Lord Woolton might be called a loaf-warden, but very few peeresses would be much use at baking today. The social changes in the next hundred years may be as great as those in the last thousand, and it would be interesting to guess which of our words of today will be regarded as specially honourable in 2042. I think 'comrade' is one of them.

If substantives change their meaning, adjectives turn somersaults. I will only instance 'jolly' and 'nice'. The Greek word 'diabolos' means originally one who throws through, as a good bowler throws through the batsman's defence. Then it meant a prosecutor, including the angel who was supposed to accuse men to God. This accuser was also supposed to be a tempter, and 'diabolos' is the root of 'devil'. A medieval Latin word for devilish was 'diabolivus'. As long as Christianity was a people's religion, the devil was regarded as an enemy. But in the Middle Ages the Church as a whole was lined up with the nobles against the people, though it produced a number of priests, like John Ball who died for freedom. Hence many people regarded the devil as a friend, and 'diabolivus' has become 'joli', or pretty, in French, and 'jolly' in English.

'Nice' is derived from the Latin 'nescius', ignorant, and meant foolish, weak, or simple, in medieval French. In Elizabethan English it meant trivial, as in Shakespeare's 'every nice offence'. Then it came to mean accurate, as in the modern word 'nicety', and still later to mean pleasant, perhaps because craftsmen enjoy accuracy.

When the history of our country is taught as the history of social change, the history of our words will be used to illustrate it, and such facts as these will be part of our general education.

4

What I require from life

I have got to accept the universe as it is. I must not require the impossible, and I shall do harm rather than good if I try to imagine perfect beings in a perfect world. But given the world as it is, I can say what I may reasonably hope both for myself and for others.

I was born in a peaceful age, and in my youth I looked forward to a life of peace. Since 1914 I have been living in a heroic age, and I see no prospect of surviving into another epoch of peace and quiet. So I must try to make the best of the time in which I live. What do I ask for myself? I assume that I have food, water, clothes, and shelter.

First, work, and a decent wage for my work. Aristotle defined happiness, not as a sum of pleasures, but as unimpeded activity. I want work which is hard but interesting, work of which I can see the fruits. I am exceptionally lucky because I can choose my own work to a large extent. If I want a respite from science I can go and be a war correspondent, or write children's stories, or make political speeches.

So I enjoy a good deal of my second requirement, freedom, in fact vastly more than most people. But I want still more, particularly more freedom of speech. I should like to say and write what I think about Lord Blank's newspapers, Mr Dash's pills, and Sir John Asterisk's beer, all of which are poisonous. The law of libel prevents me from doing so.

I require health. I don't mind an occasional toothache or headache, or even an acute illness every seven years or so. But I want to be fit for work and enjoyment in the intervals, and to die when I can work no longer.

I require friendship. Particularly I require the friendship of my colleagues and comrades in scientific and political work. I want the society of equals who will criticize me, and whom I can criticize. I cannot be friends with a person whose orders I have to obey without criticism before or after, or with one who has to obey my orders in a similar way. And I find friendship with people much richer or poorer than myself very difficult.

These four things are general human needs. For myself I also demand adventure. Life without danger would be like beef without mustard. But since my life is useful it would be wrong to risk it for the mere sake of risk, as by mountaineering or motor racing. As a physiologist I can try experiments on myself, and I can also participate in wars and revolutions of which I approve. By the way, love of adventure does not mean love of thrills. I spent six weeks in Madrid during the recent siege. The only thrill that I got there was from reading Rimbaud's poetry. The satisfaction of adventure is something much more solid than a thrill.

There are other things which I desire, but do not demand. I like to have a room of my own with some books, good tobacco, a motor-car, and a daily bath. I should like to have a garden, a bathing-pool, a beach, or a river within easy reach. But I have not, and I bear up quite happily.

I am an exceptionally lucky person because I get a good deal of what I want, and can work actively for the rest. But most of my fellows do not enjoy what I regard as essential requirements. And I cannot be completely happy while they are unhappy.

I want to see every healthy man and woman on the planet at work. But everywhere outside the Soviet Union there is unemployment, though very little in Sweden. I am a socialist because unemployment, at least during times of depression, is an essential feature of capitalism. I want the workers to see the fruit of their own work not in profits for others, but in their own and their friends' well-being. My main personal complaint is that my work is not applied. I discover new biological facts, but no use is made of them, because although the community would benefit, no individuals would make profits from their application.

I want to see the workers controlling their conditions of work as I control my own to a considerable extent. Most work is dull, much of it is unhealthy and exhausting. This need not be the case, and I belive will not be after a few generations of democracy in industry. How pleasant work can be is shown by a simple fact. When we have time and money to spare, two of our favourite occupations are hunting and gardening, the work of our paleolithic and neolithic ancestors respectively. I am a socialist because I want industry to be controlled by the workers. Freedom should begin in the workshop.

I want to see every man and woman as healthy as possible. This implies food, housing, and medical attendance of the quantity and quality which human biology demands and modern technique can supply.

I want to see the end of class subjection and sex subjection. Only so will the equality which is the condition for fraternity be achieved. Since the main barriers between classes and the main reasons for the subjection of women are economic, I look to a revolution in the economic field for their end.

I am a socialist because I want to see my fellow men and women enjoying the advantages which I enjoy myself. I know that socialism will not confer all these advantages in an instant, but if I live to see capitalism overthrown and the workers in power through most of Europe I shall die happy.

Certain things are lacking in my list of requirements, notably peace and security. It is futile to require things which one is most unlikely to obtain. Fascism is a living reality, and fascism, as Hitler and Mussolini explicitly state, and prove by their actions, implies war. War is spreading at present. I sincerely hope that it will not spread over the world, as it spread from 1914 to 1917; but I do not look forward to perfect peace till fascism is dead.

I fully realize that peace and security are rightful aims, and that my own desire for violent adventure is probably merely an adaptation to the age in which I live. I am a child of my age, and all the worse for being one. I therefore demand security rather than adventure for others.

I have said nothing about many things which I desire to see, such as a spread of education, and an increasing application of scientific methods in all branches of life. From what I have seen in Russia and in Spain I do not doubt that these and other good things would follow almost automatically if our class distinctions were abolished.

To sum up, for myself I require food, warmth, work, liberty, health, and friendship. For the society in which I live I require socialism.

Supplementary to my requirements of life are my requirements of death. Of all men whose deaths are recorded, I consider that Socrates' was the most enviable. He died for his convictions, when he could easily have survived by betraying them. He died at the age of about seventy, still in full possession of his faculties, but having completed all the work which he could reasonably hope to do. And he died laughing. His last words were a joke.

I do not require of death that I shall be as fortunate as Socrates. A death which fulfils all the three conditions of his is very rare. But if I can achieve even two of them I shall have done well, and though my friends may lament me, I trust that they will not pity me.

5

What scientists in Russia are doing

I have just received a batch of numbers of the Doklady, or proceedings, of the Soviet Academy of Sciences for the latter part of 1945. They are short papers, in many cases summaries of work which will be published more fully elsewhere. In this respect the Soviet Academy agrees with the French and American Academies of Science, and differs from our Royal Society, which publishes fewer but longer papers in its Proceedings. The Soviet method has the advantage that it enables a reader to get a bird's-eye view of the progress of research. And as the Doklady are translated into English and French they are particularly valuable to foreigners.

No one who takes the trouble even to look at the Doklady can possibly accept the widely propagated view that Soviet scientists are compelled to spend all their time on problems of immediate practical importance and to neglect fundamental questions.

*

Two of the most eminent mathematicians, Vinogradoff and Linnik, have worked mainly on the theory of whole numbers, for example the distribution of prime numbers such as 7, 17, 37, 47, 67, 97, etc., in arithmetic series such as 7, 17, 27, 37, 47, 57, etc.

Again Levin has worked on records of the height and speed of 384 meteors, and agrees with American astronomers that none were moving so fast that they must have come from outside the solar system. On the other hand, he disagrees with Lord Cherwell and Dobson about the density of the upper air, in which they flare up.

Equally far from practical application were the remarkable investigations of Promptoff and Lukina on the language of birds. They worked especially with the great titmouse, which has 20 distinct cries which are uttered in different situations, and understood by other birds. When tits were hatched and reared in the laboratory, they only got three 'words' right, the shriek of fright, the long-distance call, and the fighting shriek.

They did not understand others, for example those used when food is found, when birds are gathering into a flock, and above all those used in lovemaking and nesting. And if they uttered these cries, they did so at the wrong time. In fact the noises made by such birds have social significance, and their meanings must be learned. They can fairly be called a rudimentary language.

A great deal of very careful work on animal behaviour is being done; thus, to take an example at random, Sheglova found that field mice which were allowed a choice of temperatures chose a cooler place to nest when pregnant than otherwise.

It is extremely hard to do work of this sort unless one is genuinely fond of the animals concerned.

Along with this 'pure' science, a great deal was published which can obviously be applied fairly quickly.

An immense amount of geological work is constantly being published. For example, on the same day, Vialov described the classification of the cretaceous rocks of Ferghana in Central Asia, and Teodorovich the classification of the carboniferous rocks of Bashkiria. In each case the beds were distinguished by characteristic fossils particularly of shellfish, and attempts were made to correlate them with rocks in Western Europe. Thus clays and sands were being laid down in shallow water in Farghana, while chalk was formed in deeper water in what is now England; but limestone was formed in deep water in Bashkiria while our coal measures were being formed in swamps.

No particularly valuable minerals were reported in either region, but the more is known of the geological structure of a country the more readily are valuable minerals found and the better it is possible to site new roads, villages, and so on, and to plan its agricultural development.

An immense amount is published on mineral chemistry, both as to the exact composition of various rocks and more particularly the processes by which they are formed and changed. Indeed, this is one of the branches of science in which the Soviet Union leads the world.

A good deal of the mathematics, though abstract, arises from very concrete problems. Kolmogoroff and his pupils are dealing with the foundations of the theory of probability, work which is badly needed for State planning. If you are planning for a sixth of the world, you must try to ensure, for example, against railway congestion by seeing that the same line is not wanted at the same moment for the transport of wheat, cotton, and holiday-makers. You must form reserves to deal with possibilities, such as a simultaneous frost in the Volga basin and drought in Uzbekistan.

In capitalist countries great trusts, such as the Bell Telephone Corporation in the United States, have had to devote a good deal of effort to the study of probability, but the need is naturally greater where planning is on a greater scale.

Problems of artificial radioactivity are being tackled. For example, Mescheryakov wrote on the absorption of fast neutrons by heavy nuclei, and the resulting radioactivity.

Finally, a great deal of published work arises immediately out of practical problems. Some of it was clearly secret during the war. For example, even before the war was quite over, Wolkenstein published work on the use of coloured lights in blackout lighting. Blue lights are recommended for marking roads and obstacles, but yellow lights for working, in order to give the best results on the ground without being seen from above.

Other practical problems have nothing to do with the war. Thus Kovda describes the fight against salt accumulation in the soil of oases.

All Britain drains into the sea, so though there is a little salt in our rivers it does not stay in the soil. But where a river does not reach the sea or a basin such as the Caspian, this salt accumulates where the water dries up. So unplanned or badly planned irrigation may ruin the soil in a few years. And special measures are being taken particularly in Uzbekistan and Kazakhstan, based on systematic water analyses, to keep the salt at such a depth in the soil that it does not damage crop plants. At the same time Krupenikoff has studied local races of trees, including pines, poplars, and hawthorns, which can grow in salty soils where other trees are poisoned.

These are only a few examples taken at random from a great mass of published articles. None of them are accounts of epoch-making discoveries. But they give an idea of the very wide range of work which is being done, and the balance between practice and theory, which characterizes a planned campaign of scientific research.

6

How we can date the past

Until a generation ago the only events which we could date accurately were those which were part of a history where each year had a numerical date, such as the year 412 after the founding of the city of Rome, or a date by names of officials, such as the year when Caesar and Bibulus were consuls.

Even where the length of kings' reigns were accurately recorded, one does not know if 'ten years' means ten years and one day or ten years and three hundred days, so the uncertainty soon piles up.

About 30 years ago, a number of historic eclipses were accurately dated, so that we know the date of the siege of Troy within ten years, and, what is more remarkable, that the city of Ur in Iraq was destroyed by the Elamites in 2283 BC. This is still the earliest dated event in human history.

Two methods

Where there is no written history, there are two methods by which we can give an exact date to buildings or tools which tell us a good deal more about men than lists of kings and battles.

One method is by tree rings. In a country like Arizona, where there are so many dry years that the trees are growing under a severe handicap, far more wood is formed in a wet than a dry year. By examining recently cut trees one gets a calendar for several hundred years back.

Doughlass, the pioneer in this investigation was not content with this. He examined tree trunks from abandoned 'Indian' buildings which had been preserved in the dry climate. He found some whose ring pattern overlapped that of living trees. That is to say their outer layers showed the same succession of broad and narrow rings as the insides of trees recently felled. From these he worked still further back in the same way, until the earliest date determinable is about AD 400. Thus he was able not only to date house and tools, making a cultural history possible, but other workers by examining giant Sequoia trees have obtained at least a rough history of the weather of California for the last 3,000 years.

Mud layers

In wetter countries, annual layers of mud are laid down in some lakes. When heavy rain or a big thaw brings down a lot of mud, the layer is thick. In this way different mud sections can be compared and finally synchronized. And where human tools are embedded in the mud they can be dated.

This method of dating by annual mud layers works best in the neighbourhood of retreating ice sheets. It has been particularly used by de Geer to date mud layers or varves, as they are called in Sweden.

But, unfortunately, men were rare or absent in the neighbourhood of these giant glaciers, so such records can only be used rather indirectly to date human events. Still they do give us the earliest dated event.

In the year 7912 BC the fresh water lake occupying the northern Baltic basin, but considerably above the present sea level, burst across southern Sweden in Vestergotland, opening a valley which lowered the lake level by 90 feet. The present connection further south was only opened later, when the sea level had been raised by melting ice.

We have a pretty good record of successive events during the ice ages. For example, it is quite clear that there were four main cold periods with warm ones between them. But whether the last one went on for 50 or 100,000 years is not so certain. It may be certain a generation hence.

The Yugoslav scientist Milankovitch claims to have dated the ice ages on the basis of an astronomical theory. At present most geologists do not accept the calculations. This does not mean that they think he is wrong, merely that they are not convinced. In another 20 years they may be convinced.

If Milankovitch is right, we can date the big climatic changes of the last half million years, with errors which vary between 5 per cent and 20 per cent.

The ice ages began about 600,000 years ago according to his calculations. Other estimates give over 700,000 years.

Incomplete

The method of annual mud layers can be used to date sections of the past. Sometimes the evidence for a yearly cycle is overwhelming. For example, fossils of adult insecta are found in one part of each layer, and of their larvae in another.

Bradley counted enough varves in the Eocene formations of Colorado, Utah, and Wyoming to be able to say that a particular epoch lasted between five and eight million years. During this time a thickness of about 2,000 feet of sandstone and oil shales was laid down in two lake beds. But this only gives us the duration of about a third of the Eocene period, or a tenth of all the time which has gone by since these rocks began to form. Other estimates are based on the rate of animal evolution.

The best dates for remote events are given by radioactivity.

If a rock contains uranium, just one atom in 6,578 million of this element is transformed every year.

Most of them pass through the stage of being radium, a few through that of being actinium, and in each case end up as lead. A very tiny fraction split in two, as in atom bombs.

Thorium, another radioactive element, behaves in a very similar way.

Thus, if a mineral is found containing uranium, but no thorium or lead, one could calculate its age from the amount of lead contained in it. However, this would be almost useless in practice, because one could never be sure that some lead had not been present in the original mineral.

Fortunately, however, the lead derived from radioactive atoms has a different atomic weight from ordinary lead. So it is possible to find out how much of the lead in a mineral was derived from the uranium or thorium in it, and thus estimate its age exactly.

On this basis enough rocks have been dated to fix the dates of all rocks carrying fossils with an error which rarely reaches 10 per cent. We thus have at last the time scale of evoluation.

We know that about 600 million years have gone by since the earliest rocks were formed, containing well-preserved animal remains in any quantity, and 400 million since the first record of the vertebrates.

Two hundred and seventy million years ago our ancestors left the water, and 70 million years ago the mammals took over from the giant reptiles and became the dominant land animals.

On trial

The complete story of how these figures were reached is, of course, a very long but extremely interesting one.

Any good public library should contain Zeuner's *Dating the Past's* and Holmes. *The Age of the Earth*, and Zeuner's series of clues make even the best detective stories look pretty elementary.

Most of his ideas are generally accepted: however, Milank Milankovitch's theory is still on trial.

And in 1941 Holmes, as a result of more lead analyses, has arrived at 3,000 million years for the age of the earth, where as the oldest known rocks are little more than half this age.

7

Averages

You cannot get far in economics or politics without using statistics. And you cannot get far in statistics without using an average; for example, the average wage of a woman in the engineering industry, or the average age at death of an anthracite miner.

Now an average has two functions. First of all, it is representative. It gives one an idea of the size of a fairly typical member of the group and makes a comparison between groups possible.

One can say at once that, on the whole, women in the British engineering industry get less than men, but a very great deal more than men in the Indian industry. Similarly anthracite miners on the whole do not live so long as other coal miners, but live longer than slate quarriers.

Secondly, the average tells you exactly what share each individual would get of money, land, or any commodity which can be divided up, if it were shared out equally. It is much easier to understand that the national income is £100 each, on average, for 40,000,000 people, than that it is £4,000,000,000.

Now, when you know the average, you can answer the second question exactly. But the average is not necessarily representative.

Suppose there are two villages each with twenty families. In the first village one family has an income of £10,000 per year, and nineteen have an income of £150. In the second village ten families have an income of £300 and ten of £200.

Then the average family income in the first village is £642 10s., in the second it is £250. Yet on the whole the people in the second village are richer. If you pick a family at random from each village it is nineteen to one that the family from the second village will be richer. We can only say that the first village would be richer if the incomes were divided evenly; but that is a big 'if'.

How can we choose a representative value so as to get over fallacies of this kind? Instead of the average we choose what is called the median. If you stand 101 men in a row in order of tallness and then measure the height of the middle man, you get a height which is often more representative than the average. There are as many taller than the median height as there are shorter than it.

In the first village the median income would be £150, in the second between £200 and £300.

Another advantage of the median is that it enables you to give a representative value for qualities which cannot be measured, but which can be put in order.

I am probably less musical than most of my readers, and better at mathematics. But there is no sense in saying that I am half as musical as you, or three times as mathematical.

But we could apply simple tests of musical ability to a hundred people taken at random, and I would probably be in the worst ten. One could take the middle person in such a sample as a representative.

Sometimes, but not always, the mean, or average, is roughly equal to the median. This is so for heights of human adults, but it is not so for weights.

The average weight of Englishmen about 1880 was 156 lbs. But the median was only 147 lbs. This is because in a sample of 8,000 men there were some very fat ones.

Ten weighed over 250 lbs, whereas none weighed under 90 lbs. So the fat men helped up the average, but each counted no more for the median than if they had weighed only, say, 200 lbs.

In the same way very rich men have a large effect on the average income, but very little on the median. If we keep these distinctions in mind, we shall avoid being taken in by arguments about averages.

When we have determined our average, or our median, value we next want to know something about the spread round it. Clearly, for example, the spread of human weights is bigger than the spread of heights.

Quite a number of people weigh half as much again as the average, i. e., over 234 lbs. But the average male height is about 5 ft 8 ins, and a man who measures 8 ft 4 ins can make a living as a giant in a circus.

Probably the best measure of the spread is by means of what are called the quartiles. If we measure the 26th man from the top and bottom in our row of 101 men arranged according to height, they will give us the upper and lower quartiles.

One-quarter of the heights exceed the upper quartile, one-quarter fall below the lower quartile. In this case the quartiles are about two inches above and below the median.

The difference between quartiles, divided by the median, gives a fair idea of the spread. This is only about 6 per cent for heights, but probably over 100 per cent for incomes of English adults.

The total range does not give such a good idea. A sample of a thousand people might happen to include a dwarf or a giant, but probably would not. So the range which includes the middle half or the middle four-fifths of the population is more useful.

There is only one case where the extremes matter, and that is when they are socially valuable. One Newton is worth a thousand school teachers with mathematical degrees, one Beethoven worth a thousand men who can improvise on the piano. It may be that in such cases a high range of attainments is more important than a high average.

Of course, statisticians have devised a number of other ways of picking out a representative, on the one hand, and measuring the spread round it on the other. But if we understand the meaning of an average, and the false conclusions which can be reached by using it wrongly, we shall have made a good start in understanding statistics.

8

Quantity and quality

Students of Marxism often find the principle of the change of quantity into quality difficult to understand. And opponents of socialism never seem to realize its existence. Some of them say that under socialism no one would own any private property, even a pair of trousers. Others claim that the Soviet Union is not truly socialist because workers can lend their savings to the state, and draw interest on them.

We can understand the fallacy in such statements if we take some examples from physiology. If these critics were consistent they would go in mortal terror at every breath, because the nitrogen and oxygen of which air consists are deadly poisons—if you have enough of them.

About one-fifth of the air consists of oxygen. We use about half a cubic foot of this gas per hour at rest, and four cubic feet during very hard work. If we breathe any gas, such as nitrogen or hydrogen, which has no oxygen mixed with it, we become unconscious in less than a minute, and die within five minutes. Oxygen is an absolute necessity of human life. Luckily it is so common that nobody has been able to monopolize it.

However, as we go up the air gets thinner. At about 19,000 feet there is only half as much air in a cubic foot as at sea level. Mountaineers can acclimatize themselves to live at this height. But if one goes up to it quickly, as in an aeroplane, one becomes silly at once, and quite ill after a few hours. These symptoms are at once relieved by breathing pure oxygen, or even air to which a fifth of its volume of oxygen has been added. So the crews of aeroplanes need oxygen, and various firms make quite a good thing out of their need.

Oxygen is also used for treating some lung and heart diseases at ground level. But yet it is a poison. Pure oxygen at ground level is not very poisonous, though if one breathes it for two or three days it causes inflammation of the lungs. But at high pressures it is a violent poison. A diver sixty-six feet below the sea is under a pressure of three atmospheres. Before air can be pumped down to him it must be squeezed into one-third of the volume which it occupies at sea level. This is what we are doing, incidentally, when we fill a tyre at 30 lb per square inch pressure in addition to the 15 lb pressure of ordinary air. It would be very convenient if we could give the diver pure oxygen to breathe. If so he could come up without waiting, as he would be in no danger from the formation of bubbles of nitrogen in his tissues, which may cause severe pains, called bends, and paralysis.

Behnke and other American scientists have found that at this pressure oxygen affects the brain, and above all, the eyes, so that after three hours a man becomes almost blind. He can only see things straight in front of him, and even then not very clearly. He cannot see sideways at all. Luckily he recovers in a few minutes. When oxygen is breathed at four atmospheres' pressure, Behnke found that convulsions came on after about forty minutes; and very unpleasant they are. Even before this time cramp is said to develop in a muscle which is working hard. At higher pressures oxygen causes convulsions still quicker.

Some organisms are killed by the oxygen of ordinary air. Among them is the bacillus which causes lock-jaw. You will not get lock-jaw from rubbing earth into a scratch. But you may get it if earth containing the spores of the bacillus is carried into a deep wound where oxygen cannot penetrate.

Nitrogen is also poisonous if you get enough of it. If you breathe air at ten atmospheres' pressure, corresponding to a depth of 300 feet, you very soon become rather silly. Divers at this depth often cannot carry out instructions properly, or do skilled work, and the American workers have made it almost sure that this is due to the nitrogen in the air breathed. I have produced further evidence myself.

Finally, water is a poison. If enough water gets into your lungs you are, of course, drowned, but this is not what I mean. You can be poisoned by drinking too much water as surely as by drinking too much beer or whisky. A normal man cannot be poisoned in this way, because he excretes unwanted water with his kidneys. But this excretion can be temporarily prevented by injecting one of the hormones from the pituitary gland; and if this is done, two or three gallons of water will give you convulsions or cramp not unlike those of oxygen poisoning. They can at once be relieved by injecting strong salt solution into a vein, which brings the composition of the blood back to nearly normal.

In the same way everyone knows that you can have too much or too little food, heat, light, and other good things. Aristotle and other Greek philosophers applied the same principle to social occurrences. Aristotle said, for example, that the coward took too few risks, the rash man too many, and the brave man the right amount. And Marx constantly used the principle in his economic arguments. He showed, for example, that a large sum of money could be used as capital, but a small sum could not. No doubt when socialism has developed into communism there will be no such thing as individual savings, for one thing because there will be no need for them. For everyone will get not merely necessities, but many things which we now regard as luxuries, free.

And under communism there will doubtless be some private property. If we understand how quantity is transformed into quality, we shall realize that private property, like oxygen, can be both a necessity, as in the case of boots, and a public danger, as in the case of armament shares. And we shall steer our way between the extremists of the left, who think that a Soviet worker is a capitalist because he lends a few hundred roubles to the state, and those of the right, who think that because I can own a fountain pen, the Duke of Westminster should be allowed to own hundreds of acres of London.

9

Biology and town-planning

The Editor of the *Star* has asked me how, as a biologist, I should plan London. My first impulse was to refuse to write on the subject. For I certainly don't know enough about human biology to answer his question. Nor does anyone else. And sometimes it is worth while telling the world how little we know.

If a government asked me to plan the feeding of London's children I should take the job on. For I know where to go for figures as to what food they need, and what it costs. But there is no biological standard of housing today any more than there was a biological standard of diet fifty years ago.

We know a few basic facts about the effect of environment on health. Or rather we know what the position was in 1921. Since then our government has published very little on the question, because no one appears to be interested in it. Labour and Liberal men and women find it just as boring as Tories, because what little knowledge we have is based on statistics.

We know that country life is far healthier than town life, and that poverty is not a serious cause of death in the country, since farm labourers live as long as their employers, except for a slightly greater infantile death rate. We know that in the towns the poor die very much quicker than the rich. And most of the occupations which are healthy for town dwellers, such as driving vehicles, involve a good deal of exposure to fresh air.

We don't know how much of the high death-rate of the urban poor is due to overcrowding. Certainly not all. Some is due to the fact that most of the people who can't afford a proper house can't afford proper food either.

And there are more sorts of overcrowding than one. Mr Smith lives in a nice roomy house at Edgware, but spends an hour a day in grossly overcrowded tube trains. He might be healthier if he lived in a much smaller house in central London, but walked to and from his work. Mr Jones has a roomy house, and walks to his work, but he goes to a stuffy cinema every evening, except when he attends rather crowded political meetings where I and other speakers spray him with germs from our throats. Should we plan to avoid these varieties of overcrowding, and would anyone thank us if we did?

I pass over the more obvious aspects of planning, such as broad straight roads, zoning of separate residential and factory areas, parking places for motor vehicles, and so on. Even with considerably better roads, and staggering of hours of work, we should still have gross overcrowding of transport if the present tendency to work in the centre of London and to live on its edge, continues.

So my first task would be to try to make central London a place fit to live in, with open spaces for children to play, and as much sunlight as possible for everyone. We can only make room for the present population if we build taller buildings. These buildings would take the form of large blocks of flats. The individual house in a large town is a luxury. Only millionaires can afford one in central New York, and as millionaires and planning don't go together, no one would be able to afford one in a planned London.

There would be a minimum standard of crowding, and a maximum, to be gradually approached for all, of a small bedroom with its own shower-bath and lavatory, for each unmarried person over three years old. The number of living-rooms per family would rise from a minimum of two. Cooking would be largely communal, but I see no reason to discourage meals in the home.

One serious problem arises in the planning of large blocks. If every room is to have an outlook onto a street or garden the buildings would have to be so tall and narrow as to keep most of the sun from the ground level. I do not believe in the need of direct sunlight for bedrooms. And fresh air is somewhat of a fetish. If enough people sleep in one room it must certainly have its windows open if they are not to infect one another. But ventilation with purified and conditioned air is probably as healthy, and certainly cleaner, than ventilation by open windows. So in the interest of light for London as a whole I would be willing to have many bedrooms, but no sitting-rooms, opening on narrow walls, and artificially lit.

Every roof would be a garden, and about one block in four would be an open space. Smoky fires and steam locomotives would be forbidden, and buildings faced, so far as possible, in white or bright colours. Even so we should have enough dull and rainy days to make provision for indoor physical recreation necessary. Here I should be inclined to copy the Peckham health centre on a very large scale. This really admirable institution provides swimming and other exercise for its members, and also facilities for a medical examination from time to time even when they are well, and whenever they are ill.

London would have a thousand of these health centres instead of one, and the emphasis in the medical profession would pass from the cure of disease to the positive encouragement of health. The general public would be far more willing to undergo an annual medical inspection if it were conducted in a building which was also a clubhouse.

Among the most overcrowded places are public-houses. This is partly because there are not enough of them in some parts of London, but mainly because there is nowhere else where a working man can go of an evening to talk. My plan would include a great variety of alternatives. A group of men or women would find it easy to hire anything from a small room to a large hall for any communal purpose, music, chess, politics, drama, or simply conversation.

Some would serve a double purpose. For example, if most of the young married women went to work, as they would in a planned London, we should need plenty of crêches in which they could dump their babies during the day. These could be used by workers for recreation in the evening.

It would be up to planners to break the monotony of a planned city in such ways as this. There would be other planned recreations on a larger scale. I would have at least half a dozen zoos in London, and several botanic gardens besides Kew. They would be part of a general scheme to get Londoners to think in terms of life rather than mechanism.

A very important part of the plan would be experiment. We know something of the effect of hot wet air in the factory on human health from its killing effect in the Lancashire cotton mills. We know very little about the healthiest temperature in the home. Yet if climate out of doors affects health, so should climate indoors. Different buildings would have their air supply at different temperatures and the effects noted. The health of people living on the north and south sides of a building would be compared. Only so would we get the needed data for a further step in planning

Would this make us a nation of valetudinarians? I don't think so. I am not a miser because I look at my bank balance once a month. In a socialist community once a general level of comfort was reached, a noncompetitive struggle for health might well supersede the competitive struggle for wealth.

10

Inventions that made men free

No one doubts that the great inventions of the last two centuries have revolutionized human society, and profoundly altered the course of history. To that extent everyone is a Marxist. However, opponents of Marxism go on to say that these inventions depended on the development of scientific theory, and that the really revolutionary influence has been that of scientific ideas. There is some truth in this; but only in some kinds of society does theory lead to invention, and it is worth while for Marxists to know something of inventions which were certainly not based on any scientific theories, and which changed the course of history.

After the western part of the Roman Empire collapsed in the fifth century AD its territories were occupied by various 'barbarian' nations, such as the Angles and Saxons in England, the Franks and Burgundians in France, the east Goths and Lombards in Italy. They were uneducated, but they did not practise an economy based on slavery, and the ordinary man in Europe was probably a good deal freer and nearly as comfortable as his ancestors had been under Rome.

Literature, science, architecture, and so on, were at a very low ebb, but a number of inventions were made which had a great effect on society. Our knowledge of them is largely due to a French cavalry officer called Lefebre de Noëttes, who studied all the pictures and statues of horses, and remnants of harness, dating from more than six hundred years or so ago.

If you look at a picture of a Roman chariot, you find that it was not much bigger than a perambulator, and was pulled by at least four horses. If you look more carefully, you can see why so many were needed. Instead of having harness of a modern type, they pulled it by pressing on a strap in front of their throats. If they had exercised a force of more than a few pounds they would have been throttled.

Further, the Romans did not use iron horse shoes. They used leather ones, or none at all. So the hooves of their horses wore out on paved roads, and would have done so on macadamized or concrete roads. Horses were mainly used in open country. The Romans did not, and could not, use horses for pulling heavy carts. These were dragged by oxen, or by men. The huge stones used in many ancient buildings were largely transported and lifted by human power, often by that of slaves.

Before we start feeling superior to the Romans, we had better remember that in India, South Africa, and other parts of the British Empire and Commonwealth, men are still used as beasts of burden. This may be excusable in mountains or dense forests where horses cannot penetrate. But men pull rickshaws in many towns where there are quite good roads, presumably because they are cheaper than horses.

Some time in the so-called 'Dark Ages', very possibly in France, the horse collar and iron horseshoes were invented. This meant that when the Middle Ages began, horses were used for transport in a way which was quite impossible in the Roman Empire. There is one exception which proves the rule. Many of the heavy stones in Chartres Cathedral were drawn by teams of men who undertook this hard work deliberately as a penance for their sins. But this was unusual; and the greater use of horses was one reason why slavery was not revived in Europe in the Middle Ages.

Another invention of the Dark Ages was the rudder. The Roman ships were steered either by rowing harder on one side than the other, or by a special pair of oars on each side near the stern, one of which was dipped into the water as required. Even a sailing ship had this pair of oars. This sort of steering may have been all very well in a calm sea with a ship on an even keel. It must have been hopeless when the ship began to roll. Some ships with steering oars were still built at the time of the early crusades, but about this time the rudder superseded them, which made sailing very much easier.

However, it took a long time before wind power completely ousted man power for moving ships. The Spanish Armada which attacked England in 1588 still included eight galleys and galleasses rowed by slaves, along with 124 sailing ships; though the English had given up galleys two hundred years or so earlier, for there were none in the fleet in which Henry V sailed to the battle of Agincourt.

The first water-mill recorded in Europe was built on the Moselle under the late Roman Empire. The windmill seems to have been another invention of the 'Dark Ages', for windmills were being used in several parts of Europe by the twelfth century. The Roman mills had been worked by slaves or donkeys.

Here, then, were a series of most important inventions, all of which served to free men from the most arduous and unskilled work, namely pulling carts and oars, and turning mills. They were made in an unscientific age, and were of great political importance because they abolished completely unskilled occupations in which men were merely sources of power. In fact they were much bigger steps towards human freedom than Magna Carta, the Habeas Corpus Act, or other laws of which we learn at school. Christianity played some part in abolishing slavery, but not a very large one, for both catholics and protestants enslaved negroes. Technology was more important.

One other important invention was made in the 'Dark Ages', namely the mechanical clock. This was almost certainly invented in a monastery, while the others were made out of doors. The clock is of great historical importance as the forerunner of every kind of machine in which one wheel transmits power to another, whether through gears, belts, cranks, or worms. In the long run, therefore, it had a very great effect in liberating men from toil. But except in so far as the principle of gearing was used in wind and water mills, its effect as an agent of freedom did not show up for a thousand years.

Technology is to be an important part of education under the new Education Act. If it is properly taught, it can be made the foundation of historical teaching. If it is badly taught, it will be divorced from culture. Every teacher who is even slightly influenced by Marxism should be able to show how human progress has depended on technological improvements.

11

Astronomy

In the series of articles of which this is the first, I propose to take my readers behind the scenes, so to speak, in science, and describe some of the problems which must interest scientists today, the problems which, when they are solved, will form the basis of the science of tomorrow.

Research on them is going on in hundreds of laboratories and in the field. The results are published in scientific journals. But they are generally incomplete. And, what is more, the most important work is often the least showy. For it consists in methods. A method for covering a glass mirror really evenly with a thin layer of aluminium may do more for astronomy than the building of a giant telescope. A method for estimating the amount of fat in a drop of human blood may render a scientific investigation of fatness possible.

The plain man complains that scientists are unpractical people concerned with useless problems like counting the hairs on beetles or analysing the light from distant stars, when they should be finding out how to cure colds or make cheap bomb-proof shelters. There is some truth in this criticism, but we scientists have three excuses.

Firstly, science is so badly organized that many of us have no chance to tackle practical problems. Those who work for firms often get no opportunity of investigating fundamental problems; and, owing to industrial secrecy, the workers in university laboratories have too little chance of tackling practical problems. In the Soviet Union things are better. The same laboratories are engaged on problems which they set themselves and problems set them by the directors of industry.

Secondly, a lot of research is done as the result of tradition. A century ago the exploitation of the colonies was a crude process, and the classification of plants and animals very important. It was necessary to know whether a particular tree was teak or a bush tea. Today economically important plants are grown in plantations. But many people are still paid to classify plants and far fewer to breed them.

Thirdly, the very best research is often quite useless at the time, but immensely important half a century later. Faraday discovered electromagnetic induction in 1831, but electric power was first generated for sale about 1880. In just the same way some people thought Marx was wasting his time in drawing distinctions between labour and labour power, or constant and variable capital, when he might have spent all of it in trying to raise wages. But his theoretical work has borne fruit a thousand-fold in practice. In these articles I am going to deal mainly with research which is now useless, but may become very important.

Let us begin with astronomy, the oldest of the sciences. The main work is done on stars and nebulae beyond our solar system. Most of this work consists of accurate measurement. For example, all the stars are moving, but most of them are too far away for us to detect their motion directly. So, if a star can be detected moving relative to the others in the same direction—that is to say, appearing in the same part of the sky—it is probably close.

Also the nearer stars seem to move against the background of the further ones each year, owing to the earth's motion round the sun, just as near objects seem to move against the background of far ones if we first shut one eye and then the other. So the modern astronomer can go to bed at night, leaving an assistant to mind the telescope while a photograph is being taken.

But when the plate is developed, he will spend a day comparing it with another one of the same stars taken some years back. And to measure distances between images of stars he uses a microscope. The apparent motions of the brighter stars have all been measured. Some, like Sirius, are bright because they are near, though they do not give out much more light than our own sun. Others, such as Arcturus, are a long way off, but appear bright because they give so much light.

It is quite sure that we know all the stars which give out as much light as our sun, and are within ten light-years of it—that is to say, so near that their light does not take more than ten years to reach us. But we have no idea how many dim ones there may be within the same range. The three dimmest of the near stars were only discovered within the last ten years, and no doubt a lot more will be found. If this is so, it may turn out that very dim stars are far commoner than they seem to be now.

In fact, our present astronomical ideas may be as misleading as history, which is mainly concerned with kings and generals and not with plain men and women. And the dim stars, like the ordinary people, are in many ways the more interesting. For the bright stars consist of gas. But some of the dim ones are much denser than anything known elsewhere, and enable us to study matter in a state which we cannot yet copy in our laboratories.

Other astronomers study the chemical make-up of the stars by means of the spectroscope, which reveals the gases in their atmospheres. It might be thought that this was a very impractical activity. But helium, the lightest noninflammable gas, was discovered in the sun before it was found on earth. Its main use is for filling airships and balloons.

Today the study of spectra is not likely to disclose new elements. But it throws a good deal of light on chemistry, in a way which I will explain in a later article. And as some of the stars are much hotter, and the nebulae at a much lower pressure than anything we find on earth, they are giving us quite valuable information. Above all, the study of stars reveals properties of matter which are well displayed under different conditions of temperature and pressure, but are very much in the background on earth.

12

The Milky Way

In an earlier article I wrote about the measurement of very large and very small lengths of time, and I dare say a lot of readers thought I was being very dogmatic, like Archbishop Ussher when he calculated from the Bible that the world had been created in 4004 BC. So I shall try to explain how one of those figures was reached.

The first thing to notice is that all of them have been calculated by several independent methods, which give roughly the same figure. Unless this is done, one may easily fall into serious error. Let us begin with the biggest figure, the rotation of the sun round the centre of the Milky Way once in about 250 million years. The first thing was to get an estimate of its size.

This was done in several steps. First the size of the earth was measured. This took a long time, but is now accurately known. Then observations of the solar system were made at the same time from different points. Every century or so,* the planet Venus passes directly between the earth and sun, and can be observed with a telescope. This event looks different from different parts of the earth, as a nearby solid object looks different when you look at it first with your left and then your right eye.

So Captain Cook was sent to observe a transit of Venus from the island of Tahiti in the Pacific Ocean in 1769, and the distance from the earth to the sun determined. On the way back he annexed Australia. If he had been born in 1898 instead of 1728 he would not now be commanding one of His Majesty's ships, as he was the son of a farm labourer and apprenticed to a haberdasher at twelve years old—in fact, he was not a gentleman.

The distance from the earth to the sun was checked in another way. It was known that the eclipses of Jupiter's moons, which can easily be observed with a small telescope, were seen before the calculated time when the earth was nearest to Jupiter, while they were late when it was measured by Fizeau and Foucault in the nineteenth century, the diameter of the earth's orbit could be calculated, and the figures agreed. Since then there have been other checks.

This gave astronomers a base-line 180 million miles long instead of a mere 7,000 or so. And by observing the same group of stars first in January and then in July, we can see the near ones apparently move against the background of the far ones. In this way the distances of hundreds of stars have been measured.

* Or so—approximately.

Once again the distances have been checked. A number of what we see as single stars are shown by a telescope to be double. That is to say they consist of a light star going round a heavy one as the earth goes round the sun, or two of much the same weight revolving round their common centre of gravity. This showed that distant stars obeyed the same law of gravitation as the earth.

But if the distance of a pair was known, the speed at right angles to the line of sight could easily be calculated, and the speed in the line of sight with only slightly more difficulty. Now, when a star is moving towards us the distance between successive waves is reduced, and it appears bluer. When it is moving away, it appears redder. So the distance in the line of sight can be measured, and once again agrees with calculation.

We can only measure the distances of the nearer stars in this way. But the stars show a great deal of regularity. When we know their distance, we can calculate how bright they would be if they were so near as the sun. It turns out that the spectrum of a star—that is to say, the colour as analysed by the spectroscope—depends on the absolute brightness.

Again, some stars vary in brightness with a regular period, which may be anything from three hours to three weeks. And here, too, there is a simple relation between the period and the absolute brightness. So we have only to measure the period of such a variable, and its apparent brightness, to calculate how far away it is.

By these two methods, and others, all of which agree pretty well, it has been possible to calculate the size of the Milky Way, and the sizes and distances of the nearer galaxies. They mostly have a spiral structure and appear to be clouds of stars spinning round their centres.

But they do not spin like solid bodies. The stars near the middle complete the circuit in a shorter time than those at the outside. So by measuring the motion of the stars in the neighbourhood of the sun, a rough calculation can be made of how long it will take the sun to go once round the centre. The answer is between 200 and 300 million years.

This is an example of the chains of observation and reasoning by which these enormous figures are reached. But astronomers are not satisfied with them. They are always trying, not merely to make their figures more accurate, but to check them, if only roughly, by independent methods.

13

Is there life on the planets?

During this month all the five planets which can be seen with the naked eye are visible. Venus is an evening star, setting about an hour and a half after the sun. Jupiter is the brightest star in the sky after Venus has set, and Saturn is to be seen to the east of Jupiter. If you go to work in the morning before daybreak, you may see Mercury and Mars in the east, rising before the sun.

What are these planets? Many primitive people thought they were gods, and controlled human destiny. Their reasons seem to have been something like this. The fertilizing flood of the Nile comes just after Sirius is first visible in the morning. So Sirius controls the Nile flood. And in the same way another star controls the lambing season and a third the wheat harvest.

But some events, such as wars and pestilences, do not happen regularly. So they must be controlled by the stars which move their position relative to the others, the planets or wanderers. This argument is no worse than many which we now hear about the causes of wars and slumps. Even two thousand years ago, though the Greeks and Romans did not usually worship the stars, they thought it blasphemous to suggest that heavenly bodies were made of the same sort of stuff as earthly things.

It had long been clear that the moon shines by reflected sunlight. When Galileo turned his telescope on Venus he saw a crescent like the moon's, which altered its shape as the planet moved. It was therefore clear that the planets are cool bodies like the earth. Copernicus' theory that the planets and the earth went round the sun made it possible to calculate their distances, and their sizes were then determined by measuring their images in a telescope. Venus and Mars turned out to be about as big as the earth, Mercury somewhat smaller, and Jupiter and Saturn much bigger.

It seemed natural to speculate that they were inhabited. But before it was possible to say whether life as we know it could exist on the planets, a lot more information was needed. And our knowledge of the planets has not increased very greatly in the last fifty years, although we have found out vastly more about the distant stars and nebulae.

The reason is interesting. If we want to know more about a distant cluster of stars, we train a telescope on it, and use very complicated machinery and a still finer human control so that the telescope follows the cluster in its apparent motion across the sky. We then take a time exposure lasting for many hours.

But we cannot photograph the surface of Mars in this way, because the planet turns on its axis about as quickly as the earth. So astronomers must rely on their eyes. And as a matter of fact some of the best observation of the planets is done by amateur astronomers, including Mr Will Hay, the comedian, and several English country clergymen, using relatively small telescopes.

It is clear that we can see the solid surface of Mars, whereas in the case of Jupiter, and probably Venus and Saturn, we can only observe the tops of clouds, which may consist of drops of liquid, or of solid dust. We can follow seasonal changes on Mars. During the winter each pole develops a white cap which doubtless consists of frost. This frost may be frozen water. But it may be solid carbon dioxide, which is used in the refrigerating industry under the name of dry ice, and is only solid at temperatures far below the freezing point of water. There are also colour changes elsewhere which may be due to vegetation.

Although we have not learned much fresh information about the surface markings of these planets in the last fifty years, they have been studied with two instruments which tell us a lot about them. A sensitive thermopile placed at the focus of a telescope gives an electric current proportional to the heat coming from a star. Reflected sunlight is not stopped by a thin layer of water. But heat from a body which is warm, but not white-hot or even red-hot, is stopped. So by putting a little water bath in front of the thermopile we can measure the temperature of the planets.

Mercury and Venus are hotter than the earth, but Venus is probably well below the boiling point of water. On the other hand Mars is colder, though at least in the daytime in its tropics ice would melt. But the visible surfaces of Jupiter and Saturn are intensely cold, though the solid surface under the clouds may be somewhat warmer, especially if there are volcanoes.

We can also use the spectroscope. When light is passed through carbon dioxide certain components of it are absorbed. Not visible light (or carbon dioxide would be coloured), but infrared light, which can be photographed on a specially sensitised plate. The same is true for other gases. So by comparing sunlight reflected by Venus with sunlight reflected from the moon, we can see that the former has passed through the equivalent of several hundred yards of pure carbon dioxide at the ordinary pressure.

And there is no oxygen or water vapour in the atmospheres of Venus or Mars, or at any rate far less than on Earth. Hence a man could only live on these planets in something like a mine rescue apparatus, and it seems to me a little unlikely that there is life of any sort on Venus. If there is life on Mars it is probably more like that of the bacteria which live without oxygen in black mud than to those of familiar animals and plants. So perhaps we had better make our own planet fit for rational beings before we colonize others.

14

Common sense about the planets

IN the development of science there is a constant and fruitful struggle going on between two tendencies. On the one hand we should like to explain everything, that is to say to find a reason for it. Geologists do not believe that it is either because of pure chance, whatever that may be, or an arbitrary act of creation, that the Thames bends to the right between Lambeth and London city. But they would equally agree that we do not yet know why it does so, and that any theory is likely to be wrong, and if generally accepted, likely to hold up the progress of geology. So although it is an excellent thing to find the right explanation, it is quite reasonable to say that we do not know the explanation, and had better for the present confine ourselves to describing things as they are.

A very good example of this fruitful strife is found in the history of our ideas about the solar system. For thousands of years men have known that they could see the sun, the moon, and five other planets. Actually you can sometimes see an eighth, Uranus, if you know where to look. Early men said there were seven planets because there were seven gods each identified with one of them. When their motions were described people asked why they moved as they did, and why there were just so many.

Kepler tried to answer the question. He thought that Mercury Venus, Earth, Mars, Jupiter, and Saturn each moved in a sphere with the sun as its centre, and the spheres were so arranged that one of the regular solids would fit in between two of the spheres. As there are only five kinds of regular solid, of which the cube is the best known, there could only be six planets. Kepler found that this theory wouldn't work, and later found the correct shape of the orbits, or very nearly so.

Newton showed why the orbits have the shape they have, and why a planet at a particular distance from the sun must move at the speed it does. But he did not try to explain why the known planets were at the distances calculated. He thought that this was due to the act of the Creator, though once created, their later motions could be explained.

Hegel explained why there were just seven planets. Unfortunately for him, a lot more have been found since, and no one takes his theory seriously. But he was quite progressive in supposing that the structure, as well as the motions, of the solar system, must have a rational explanation. An astronomer called Bode produced a 'law' or rule for the distances of the planets from the sun. But it is only very roughly true, and if it were quite true, no one has suggested a reason why it should be.

This year a bit of real progress has been made by Kuiper, of Yerkes Observatory, who points out that there is a relation between the masses of the planets and their distances, and that this also holds for the satellites of the great planets Jupiter, Saturn, and Uranus. If the planets were formed from the condensation of dust or gas whirling round the sun, or from the breaking up of a filament shot out of the sun when another star passed near it, or in several other suggested ways, then the heavier a planet, the bigger the gap should be between it and its neighbours. For a heavy planet would not only use up more of the available matter, it would attract the matter in its neighbourhood, and prevent the formation of other planets near it. There are other more complicated arguments which suggest what the relation between masses and gaps should be.

So here is what Kuiper did. He added together the weights of each pair of neighbouring planets, for example Venus and Earth, Earth and Mars, and divided by the weight of the sun. He also calculated the differences between the average distances of these planets from the sun, divided by the sums of these distances. So he got two columns of figures, and found that they fitted pretty well to a relationship derived from theory. Then he did the same for the systems of moons moving round Jupiter, Saturn, and Uranus. He found that the numbers followed the same rule.

In fact the rule which they fit is not quite what would be expected on any theory yet suggested, though one simple theory is nearly right. The rule finally arrived at is not completely accurate either. In fact only about half the distances calculated are within 10 per cent of the true values. But that is something to be going on with. Fortunately for science, we shall soon have the means of finding out whether Kuiper's rule is really universally valid. Slight irregularities in the motions of some of the so called fixed stars, which of course are distant bodies like our own sun, seem to show that they too have planets. When the distances and masses of these planets are measured, if they conform to Kuiper's rule we shall have to take it very seriously. Meanwhile it may perhaps be made more accurate, by taking account of all the planets or satellites in a system, not only immediate neighbours, and by finding a more complete theory to account for it.

The important point is that Kuiper's work justifies the view held by all Marxists, and a great many scientists who are not Marxists, that there are no arbitrary features in the universe, that is to say facts for which no rational explanation can be found. Of course at the present time we can only find a rational explanation for a tiny fraction of the known facts. And no doubt a lot of the explanations which we give will turn out to be wrong when we know more facts and think more clearly about the known ones. But we are right to ask why the earth goes round the sun once a year, and Jupiter once in eleven years, why male and female lions differ more in hair length than male and female tigers or cats, why sugar and glycerine taste sweet and alcohol does not, and so forth. We don't know the answers to these latter questions, but such work as Kuiper's makes us hope more firmly that our descendants will know them.

15

Weather

If you buy a textbook of science, or learn it at school or at a university, you will be presented with the solutions of problems. You will be told how to calculate the force exerted by a magnet, or what is the cause of typhoid fever. But if you attend a meeting of a scientific society you will hear about the unsolved problems. Its members will be discussing the forces acting on the Milky Way, or the cause of cancer.

In this series of chapters I want to take my readers behind the scenes, to let them know what sort of questions interest scientists and occupy their thoughts. Many people think that scientists should be ashamed of the very existence of these problems, that we ought to know our business by now, and that if we have so far failed, the search should be handed over to someone else, to spiritualists or herbalists, Christian Socialists or Mr Bernard Shaw. This is not quite fair, for a simple reason. Every scientific problem solved gives us a number of new ones to puzzle us. Sir Humphry Davy found that common salt was made of a soft metal, sodium, and a green gas, chlorine. A great discovery, but it meant that chemists now had to study sodium and chlorine, as well as salt. Pasteur found that many diseases were caused by bacteria. His successors were left to find out how bacteria live, what they eat, how they breed, and how best to kill them.

I am going to begin with the one scientific problem which interests every man every day—the problem of weather. We can forecast it for a day or two ahead, but that is all. We cannot tell a week ahead whether it is going to rain on Bank Holiday. Above all, we can say almost nothing about the weather twenty years or two thousand years ahead. Yet the whole future of our country depends on the weather. In the past there have been big changes in the weather, lasting over centuries and moulding human history. Between AD 450 and 1000 the weather was much warmer and drier than it is now. During that time the Northmen constantly invaded England. The North Sea was generally calm, and their voyages were easy. What is more, they rowed across the North Atlantic to Greenland, which really was green in those days, and even to Canada.

If we get a few centuries of really good weather in the future, Canada will become one of the world's greatest nations, with a population of hundreds of millions, and wheatfield stretching up to the Arctic circle. There will be no more icebergs in the North Atlantic, and all through the year ships will use the great ports which will be built round Hudson Bay.

But the weather may equally well get worse. Two thousand years ago England was so wet that the valleys were full of great swamps, and the chalk uplands of Salisbury Plain were the most habitable parts of the country.

In the past such changes of weather meant the end of great civilizations. Some scientists believe that the highest forms of civilization have only been able to flourish where the weather was not too extreme, as regards heat or cold, moisture or drought, and where it was changeable enough to be stimulating. This may have been true in the past; when people spent most of their lives out of doors. But today we English live mainly indoors, in an artificial climate of our own making, which is probably altering our national character. So an urban civilization would be able to stand up to a climatic change which might have wiped out a civilization based on agriculture.

Finally we may have a real cold spell. Everyone knows that thirty thousand years ago most of England and Canada were covered with ice. The ice moved southward with such force that blocks of rock from Norway are found scattered all over Norfolk, and the village of Great Catworth in Huntingdon is mostly built on a huge lump of chalk torn from the Lincolnshire Wolds.

A generation ago scientists talked of the great Ice Age. It was over, they said, and there was no reason to expect another. But since then rather an alarming discovery has been made. There was not one ice age, but there were four, separated by warmer periods. If there have been four, there seems no reason whatever why there should not be a fifth.

It is worth while picturing what that will mean. Most of England will be covered with ice, which will grind every vestige of civilization out of existence. A few cairns on the top of the highest hills, and the mud-filled shafts of our mines, will be all that remains to show that northern England and Scotland had ever been inhabited.

The South Coast will probably be kept free by warm currents, as it was during former ice ages. The glaciers never got south of the Thames. So southern England will be a snowy country inhabited by a few fur trappers. For the clouds which now fertilize England will go farther south, and shed their rain on what is now the Sahara Desert. And the fashionable ladies of Timbuctoo will doubtless demand real English arctic foxes. Perhaps, however, a few enterprising Saharan tourists will occasionally venture so far north, and ski among the ruins of London.

On the other side of the Atlantic the ice will come a good deal farther south. The advancing glaciers will sweep aside the proud skyscrapers of New York as a broom sweeps away autumn leaves. Canada will be as desolate as Antarctica today. Occasional bold airmen flying across it will say: 'There used to be a city called Montreal somewhere near here. But of course there are no landmarks on this ice, and we can't tell the site within ten miles'.

These events sound dramatic enough. But if they happen they will not be particularly dramatic because of the extreme slowness of geological as compared with human happenings. We do not know how quickly the ice advanced during the last ice age. We do know how slowly it retreated. When the great ice-sheets that covered Canada and Scandinavia were melting there was a flood every spring which laid down a sheet of mud. These sheets have been counted, and we know that the site of Stockholm was uncovered less than twelve thousand years ago, and that on average the ice in southern Sweden retreated at the rate of only about two hundred yards per year.

So if an ice age begins next year it will be several thousand years before England is covered with ice. The first signs would probably be a series of very snowy winters and cloudy summers. The snow on the Scottish Highlands would last late into the spring, and then throughout the year. There would be a boom in winter sports in Scotland. Within a century a few small glaciers would be forming on the northern sides of some of the Scottish mountains, and all over the world the existing glaciers would be pushing their way downwards.

But there would be few headlines. The northern lands and Switzerland would be gradually abandoned. Centuries before any town was overwhelmed by the ice it would have been quietly evacuated owing to the cold, and have become a mere name in the history books. Today no one but archaeologists would care if the desert sands covered the mounds that were once Babylon and Nineveh. If the ice ever covers London and New York again a few African or Mexican schoolmasters may bore their classes with accounts of the quaint people who once lived there. But the ordinary man and woman will probably never hear of it.

Another ice age, if it happens, will certainly not wipe out civilization. Civilization will retreat southwards, as the men of the old stone age retreated before the last icesheet. And when the ice melts after fifty or a hundred thousand years our descendants will return to re-people the waste lands.

Before the present series of ice ages there had been several earlier series, probably at least four. The last one was about two hundred million years ago, long before the age of the giant reptiles, and since then there were no ice-fields till less than a million years past. A series of ice ages is a malady from which our planet occasionally suffers, just as a man a few times in his life gets a disease which gives him several bouts of fever.

When our planet finally gets over its present chill and settles down to good weather, the ice-caps on Greenland and Antarctica will melt, and men will live in these countries. Just as explorers in the Sahara ride over the ruins of ancient towns buried beneath the sand, so arctic and antarctic explorers run their sledges over the sites of cities which their remote descendants will inhabit, sites which are now hidden under an ice-sheet which may be a mile thick.

Even if we knew the cause of the former ice ages, we might not be able to tell whether there will be another. But we do not know their cause. There are plenty of theories. Some scientists think that the sun gave less heat, others that it gave more, in consequence of which more water was evaporated from the sea, the whole air became cloudy, and so much snow fell that great glaciers spread southwards. Another theory is that there was less carbon dioxide in the air, so that the earth cooled down much quicker at night. Others again think that the earth's crust slipped on its core, so that the North Pole was in Greenland. That is only a small sample of the theories.

At this the plain man may fairly say, 'Scientists must have a nice easy time making theories about ice ages. They can't be tested till the next one comes, by which time their makers will have been dead for centuries.' Luckily there is a perfectly good answer to this criticism. No scientific theory is worth anything unless it enables us to predict something which is actually going on. Until that is done, theories are a mere game with words, and not such a good game as poetry. These theories can and will be tested. If the North Pole shifted to Greenland, Siberia must have been warmer when Europe and North America were colder. If there was less carbon dioxide in the air, plants must have grown more slowly.

We can apply this simple test to any problem which we are asked to solve. There is an old problem which has been solved, though people still sometimes ask it. 'Which came first, birds or eggs?' The answer is, 'Eggs'. Birds were descended from reptiles which laid eggs. Wherever you choose to draw the line between birds and reptiles, it is sure that the first bird hatched out of an egg.

And the answer to this problem is of real practical value. We know roughly when the first birds appeared. It was about the time when the Bath limestone and the Oxford clay were being formed on the sea-bed, long after the age of coal, but before that of chalk.

This means that if we find a fossil feather we can be sure that the rock in which it is found was laid down long after the coal measures. So there is at least a chance that there may be coal underneath it. Almost the only way that we have of dating rocks is by the fossils in them. There are still a number of people who do not believe in the theory of evolution. Scientists believe in it, not because it is an attractive theory, but because it enables them to make predictions which come true. If only the critics of evolution had the courage of their convictions they would start mining for coal in Cornwall or tin in Oxfordshire.

The different branches of science are so linked up that we cannot tell how they will affect one another. My father invented an apparatus for air analysis in order to test the air in schools, factories, and mines. If the carbon dioxide rises above a certain level this is a sign that better ventilation is needed. Just before he died he managed to make this apparatus ten times more sensitive, so that one can follow the changes in the carbon dioxide of outside air, of which there is about an ounce in fifty cubic yards.

If the theory that great weather changes are due to changes in the carbon dioxide is true, this apparatus will enable us to test it. I think the odds are against this particular theory, but it would not be at all surprising if a research on stuffy schoolrooms led to a means of predicting and perhaps even preventing an ice age.

Pasteur started out to find why certain crystals are lopsided, and ended up by discovering the bacteria which cause diseases. It is the glorious uncertainty of what our discoveries may mean for the future which makes scientific research a gamble even more fascinating than horse-racing or politics.

16

Frost

WHY are frosts so important? Why does it matter more when the temperature falls below 0° Centigrade or 32° Fahrenheit, than when it falls below 20° or −20°? The answer is because water is the commonest substance which undergoes a change of state within the temperature range where human life is possible. If we could live at such temperatures, the boiling point of water, or the temperatures at which the commoner kinds of stone melt, would be equally important.

By a change of state is meant a change between solid and liquid, or liquid and gas, more rarely a change between solid and gas, skipping the liquid state. Most pure substances change very abruptly from one state to another. The freezing point, however, does not depend only on temperature, but on pressure. Most liquids get denser when they freeze. For example, when one melts lead or solder the solid unmelted part remains at the bottom of the pot, though there is a thin skin at the top where the surface is cooled by the air. In consequence if you squeeze molten lead you tend to force it back into the solid form. In other words the melting point of lead rises with pressure.

But water behaves differently. Ice is lighter than water. That is to say water expands when it freezes. It even starts to expand a little before it freezes. This has a number of consequences. If ice were heavier than water it would fall to the bottom of ponds and of the sea. Shallow ponds would freeze solid. So would the Arctic Ocean. And very likely there would be ice at the bottom of even the tropical oceans. In fact the whole earth would probably be a lot colder; and perhaps Venus, which is a good deal warmer than the earth, would be a more suitable place for living beings.

On the other hand frosts would be less dangerous to traffic. The main reason why we slide on ice is apparently that local pressure causes it to melt, so it does not give us as solid a footing as an equally smooth surface of wood or metal.

It is not so sure whether the other menace of frost to towndwellers, namely burst pipes, would be any less serious. When the water freezes in a pipe, it bursts it by expanding, particularly if some water is trapped between two blocks of ice. We do not notice the burst until the thaw occurs. But we may be able to prevent it, not only by leaving taps dripping, but by heating up a frozen pipe the moment the water stops flowing, if we can find out where the ice has formed. If water contracted on freezing, the pipes would not burst till it thawed again, but they might do so then. I leave that question to better physicists than I.

It is not only water pipes that are smashed by frost. Wet ground, and even wet rocks, are split. The sharp-edged pebbles so produced in mountains and the Arctic are quite characteristic. So is a peculiar type of soil produced when the stones and earth on a hillside have been loosened by frost, and the soil frozen so deep that when spring comes and the surface melts the water cannot sink into it and the surface layer slides down. Geologists can recognize this type of deposit, which is called Coombe rock in southern England, where it was formed during the ice ages, and accumulated in valleys in the downs just beyond the range of the great ice sheets which covered much of the country.

When waterlogged soil, such as a Siberian tundra, freezes, it first expands, but with severe frost it contracts again, and cracks form, often with a loud bang. These fill with ice which does not melt next spring, and the cracks may grow till they are two yards across.

Frost also forms peculiar rings of stones on the surface. Apparently heat leaks out quicker through a solid stone than through soil, so ice forms under stones and lifts them to the surface. In the Arctic there are no earthworms to counteract this process. So stones are heaved up above the soil, and a further action of the frost, about which there is a good deal of dispute, arranges them in rings. These, and other peculiar types of soil due to frost, have been found buried below the ordinary soil in many European countries. During the Ice Age it seems that England and France had plenty of snow in winter, but few frosts hard enough to make deep cracks. While in central and eastern Europe conditions were more like those in Alaska today.

The most serious effect of frost is of course on living beings. They all consist mostly of water, and when this freezes they are at best in great danger. Nevertheless, some can stand freezing. Of course a fish in a block of ice is not necessarily frozen. For living stuff contains enough salts and other things dissolved in it to lower its freezing point well below that of fresh water.

Curiously enough those simple organisms, such as some plants and bacteria, which will survive freezing, are safest if they are made extremely cold, in fact brought down to liquid air temperatures. At temperatures a little below freezing point the ice molecules can still move a little, and arrange themselves in large crystals which destroy the organic structure. At very low temperatures crystals can no more grow in ice than they can in glass at ordinary temperatures.

Most of our plants have by now shed their leaves, and those which have not have special chemicals to protect them. In fact most leaves of winter evergreens such as holly and spruce are full of oil and resin which will not freeze, and contain very little water. But unfortunately these frost-resistant liquids will burn. That is why Christmas trees and holly branches can rather easily catch alight, and it is safer to decorate them with electric bulbs than with candles. I suppose that every winter these anti-freeze mixtures in leaves are responsible for a few human deaths, mostly of children, from burning. The anti-freeze compounds used in car radiators also kill a few people who drink them, particularly when they are used for adulterating cheap spirits.

In England frost is not much more than a nuisance. In the northern parts of the Soviet Union it is one of the main natural conditions to be fought. Even the principles of architecture are quite different where in a few winters' frost can heave the foundations of a house out of the ground. The year is sharply separated into two parts, during which people have very different kinds of life. A Soviet town planner must think of both, and the conquest of the Arctic so that ordinary people can be as useful and happy in winter as in summer is one of the many problems which the Soviet Union is solving.

17

Is man a machine?

The French philosopher Descartes, three centuries ago, was the first to say that men were machines. He thought that animals were nothing but machines, and men were machines guided by souls. The Greek and Roman philosophers did not have this idea, because they only knew of very simple machines, such as bows and pulleys. When such complicated machines as clocks were made the idea became much more plausible. It is obviously a useful idea, because we can ask the same sort of questions about the parts of our body as we do about the parts of a machine. What is the heart for, and how does it work? Its function is to pump blood round, as an oil pump pumps oil round in many motors. And it has valves and other parts like those of pumps. In the same way we can compare the eye with a photographic camera, the nerves with telegraph wires, the fat under the skin with the insulating material round a boiler, and so on.

The description of man as a machine is much more useful to-day than it was in Descartes' time, because in his day self-moving machines were worked by springs, like that of a clock, whereas now the best-known ones are worked by the burning of coal or petrol. The comparison between a man and an engine driven by coal or petrol is very close indeed, if we merely consider intake and output, and do not go into the details. Each requires combustible food or fuel and a large supply of air. In each case most of the food or fuel is combined with oxygen from the air and converted into carbon dioxide and water vapour.

Human energy outputs and needs are usually measured in large calories, a calorie being the amount of heat which raises the temperature of a kilogram of water by one degree Centigrade. A very inactive man can carry on under 2,000 calories per day, an ordinary worker on 3,000 or so. A coal hewer may need 5,000, and a racing cyclist will expend 10,000 though he will take several days to make it up by eating. A well-trained man doing hard work may reach a mechanical efficiency of about 20 per cent. That is to say, one-fifth of his energy output is work, and four-fifths heat. But as he can only work for part of the twenty-four hours, this average cannot be kept up. A steam or petrol engine can reach much higher values, but this is hardly a fair comparison. For the engine does not stoke itself, oil itself, do running repairs, and occasionally co-operate with another engine to make a little one. Much of our energy is spent in keeping ourselves in working order. For example, all the energy which the heart expands in pumping the blood round is ultimately converted into heat.

In these important respects, then, a man is a machine. It is worth while looking at some ways in which he is not. One of the essential features of a machine is that any part of it can be replaced by a spare part. This is only true to a small extent with men. If you try to graft a limb from one individual to another, it may heal into place, but it will die later on. On the other hand, you can graft together trees of different varieties or even different species. And, what is more striking, you can do the same with frogs if you operate on young embryos. A head from one species will grow on the body of another.

On the other hand, as the practice of transfusion shows, human blood can be used as a 'spare part', provided certain precautions are taken. Again, if a man's pancreas goes wrong in a particular way, he cannot use sugar and wastes away with diabetes. But if he is given insulin from a pig's pancreas, he recovers his health. Insulin is another spare part that we can use.

So it is more dialectical to ask the question 'How much of a machine is man?' One can then answer in detail. Part of the answer would be that, as regards the use of spare parts, man is less of a machine than a tree or a frog, but still something of a machine.

18

Breathing

Primitive man thought that breathing and life were the same thing, and this curious idea is embodied in our language. There are two Latin words for breath, *anima* and *spiritus*. Our words 'animal' and 'spiritual', which are now used with contrary meanings, both meant 'breathy' originally.

Some simple organisms can live without any sort of breathing. Most of them use oxygen and get rid of carbon dioxide. The larger land animals do this through special organs, the lungs. When we buy them at the butcher's they are called 'lights'.

We expand our lungs by moving the ribs, and by contracting a dome-shaped muscle, the diaphragm, between the chest and belly, which pushes the contents of the abdomen down, and sucks the lungs after it. Men use the diaphragm more than women. The essential of human breathing is the exchange of gases between the lungs and the outer air. If the breathing muscles are paralysed this can be done artificially for years on end, if need be, with an 'iron lung', which expands and contracts the chest artificially. It can even be done without moving the chest at all by putting a man into a steel cylinder in which the air pressure is raised and lowered through about five lb per square inch gauge pressure about 15 times a minute. When a man is put in this he soon stops breathing in the ordinary sense of the word. The expansion and compression of the air in his lungs is enough to carry out the necessary gas exchange.

The lungs are of a spongy structure with a total surface area of about 100 square yards. The membrane between blood and air is extremely thin, and is the most vulnerable part of our body. About one person in seven dies of lung diseases due to infection, dust, or both, although the lungs have a very large factor of safety.

One can live quite well with only one lung. More people die of heart disease than lung diseases. But that is because no large part of the heart is indispensable. You could not live for five minutes with only half a heart. A resting man uses about half a cubic foot of oxygen an hour, and anything up to ten times as much during very hard work. He produces a slightly less volume of carbon dioxide. When the carbon dioxide in the inspired air gets above 3 per cent the breathing gets deeper; when it reaches 6 per cent one pants heavily.

So to take a topical example, if each man in a submarine has 400 cubic feet of air, during moderate work he will produce about a cubic foot of carbon dioxide per hour, and the crew will be panting badly after 24 hours if this gas is not removed artificially.

Both the oxygen and the carbon dioxide in the blood are mostly in chemical combination. You can get about 18 to 20 volumes of oxygen and 50 volumes of carbon dioxide with a pump from 100 volumes of arterial blood. This means that a volume of blood carries round nearly as much oxygen as the same volume of air.

The oxygen is combined with the red pigment haemoglobin, and this will not take up much more even if you breathe deeper and faster than usual, or breathe pure oxygen. Nor does it take up much less if the amount of oxygen in the air is slightly cut down, as when one flies at the height of a mile; though at two miles it is better to breathe air with some extra oxygen in it.

The carbon dioxide is mostly present as sodium bicarbonate, and as various other weak acids compete with it for the sodium, your blood loses a lot if you overbreathe, and takes up a lot if you breathe air containing, say, 7 per cent of carbon dioxide.

In either case you feel pretty queer. If you sit in a chair and breathe as deep and as fast as you can for a minute or so your fingers will probably begin to tingle.

If you go on longer you will probably get cramp in your hands and feet, but some people get convulsions; and probably if every reader of the *Daily Worker* tried it, one or two of them would die.

For your body to work normally you must have just the right amount of carbon dioxide in your blood and various organs. If you like to put it that way, carbon dioxide is both a poison and a necessity of life.

The breathing is normally regulated so as to keep the carbon dioxide in the blood at a steady level. You pant if it rises above this level as a result of exercise. If it falls below it you slow down your breathing, or even stop it, until it rises to normal again.

There is a second regulating device which speeds up the breathing still further if the oxygen in the blood falls below the standard level; but this only comes into action during very hard work.

The blood coming to the brain is continually being sampled partly by cells in the brain itself, and partly by a special organ situated along with the pressure gauge on the carotid artery in the neck; and breathing is a reflex depending on the composition of the blood.

In fact, breathing is one of the many bodily activities which keep the internal environment of our cells steady, and as it is one of the easiest to study, it has thrown a good deal of light on the others. It is remarkable that this regulation goes on even when we are using our lungs for another purpose, such as speaking or singing.

Normally the regulation is quite unconscious. But if it is interfered with—for example, if we hold our breath or breathe air containing 7 per cent of carbon dioxide—we develop an air hunger which gets priority over every other sensation.

This is quite typical of how we work. Our minds pay little attention to our bodies as long as they are working normally, but attend very quickly when certain things go wrong.

Some demands are clear enough to our consciousness. We know whether we are short of air, water, or food. Unfortunately we are not so clear when we are short of some particular component of our food, such as tryptophan or vitamin A.

19

Why are you left-handed?

I have had several letters from readers asking me to write about right-handedness, left-handedness, and ambidexterity. They want me to tell them why most people are right-handed, and so on. I wish I knew. Nobody does.

But I do know enough to make it in no way surprising to me that this mild kind of asymmetry is common. For the more important molecules of which we are built up are asymmetrical.

And so are our insides. The outside of your chest and belly are (I hope) pretty symmetrical. So are the bones and muscles. But the organs inside them, the heart, lungs, liver, stomach, and so on, are very far from being so.

Atoms seem to be symmetrical. At present there is no evidence that, for example, there are 'right-handed' and 'left-handed' carbon atoms, related to one another like a pair of boots, or a boot and its image in a mirror.

But as soon as a complicated molecule is built up in which four different atoms or groups of atoms are joined to the same carbon atom, we get rights and lefts.

If the four atoms were arranged in a plane, like four streets coming into a crossroads this would not be so. If we turned such a molecule over it would correspond with its mirror image.

But, in fact, they are arranged in three dimensions, nearly like the corners of a regular tetrahedron. If a carbon atom is united to four other atoms A B C D, and if we imagine A to point upward, then the other three atoms can be arranged as

$$\begin{array}{ccc} \text{B} & & \text{B} \\ \text{C}\quad\text{D} & \text{or} & \text{D}\quad\text{C} \end{array}$$

A chemist in a laboratory generally makes the two sorts of molecule in equal numbers, but a living plant or animal generally makes only one kind.

Pasteur, who was the pioneer in this work, showed that asymmetrical molecules often make asymmetrical crystals, and always cause the plane of polarized light to rotate when it is passed through a solution containing them.

The reason why an animal or plant only uses one of the two possible kinds of asymmetrical molecule is fairly simple. To transform a molecule it needs a set of enzymes which fit it. For example, when we digest meat, starch, or cane sugar we use enzymes which fit the kind of molecules which are found in nature. The mirror images of some of these molecules have been made, and most of our enzymes won't fit them.

When Alice went through the looking glass in the story she had to eat looking-glass food. If she had done so she would have been unable to digest it, and would have starved to death!

Actually, we have enzymes which will oxidize a few of the simpler kinds of 'looking-glass' molecules, but even these can only be used as fuel, not for building new living substance. They probably get made by accident in our bodies, and are scrapped as soon as possible.

It would be possible to imagine an animal or plant which used both kinds of asymmetrical molecules equally well. But it would have to have a double set of enzymes, and a means of separating the two kinds of molecules.

Naturally enough, a great many primitive animals and plants, consisting of a single cell, are asymmetrical. Symmetry only becomes useful when an animal has to move towards an object, which it can only do if it has organs, such as eyes or nose, for apprehending it at a distance.

Even so, symmetry is not necessary for slowly moving animals such as snails, and some quickly moving ones manage well without it. The most striking example is given by the flat fish.

Have a look at a plaice or sole. Its ancestors were symmetrical, and it starts life as a nearly symmetrical larva. Then it takes to lying on its left side, one of its eyes moves round to the right, and it finally develops enough symmetry from back to front to enable it to swim fairly well.

Most of our flat fish lie on their left sides. The best-known exception is the turbot, which lies on its right. But occasionally one of these fish lies on the side unusual for its species.

Apparently symmetrical animals are often asymmetrical inside, like ourselves and are always liable to develop asymmetries. Many crabs, for example, have one of their pincers much bigger than the other; some are right-handed, some left-handed.

In development and evolution there is a constant struggle between processes making for symmetry and for asymmetry. This is, of course, only one of the conflicts in evolution, and not one of the most important ones.

At the moment, asymmetry seems to be winning a temporary victory in man, for most people are sharply righthanded, and most of the rest sharply left-handed. What is less well known, most people are right-footed. At least that is so in Sweden. Only 34 out of 194 Swedes hopped on their left foot.

One reason for left-handedness is twinning. About 5 per cent of ordinary people are left-handed. But about 16 per cent of monozygotic twins, that is to say, twins derived from a single egg, are left-handed; generally only one of a pair. One might expect a pair of such twins to be like mirror images of one another, and this seems to be fairly often so as regards handedness.

Probably some left-handed people are survivors from pairs of such twins, the other twin having died long before birth. But this is certainly not the only reason for left-handedness. As usual, I can't answer the question put to me. I can only give a little bit of the answer. But even that little bit takes one a long way into biology.

20

Domestic animals

What was the most important event in man's past? I do not say in human history, because, if I am right, it took place before any historical records were made, and even before the origin of the rather dubious legends with which history begins. At first sight one would be inclined to choose some great political transformation, such as the fall of the Roman Empire or the Russian Revolution, or else the beginning of some world-shaking system of ideas, such as the origin of Christianity or the writing of the Communist Manifesto.

However, if the Marxist account of history is correct, these events, or something like them, were bound to have happened. The Roman Empire was based on slavery, and was bound to break down or change very greatly when slavery ceased to work. The New Testament is the only set of books surviving from the Greco-Roman civilization which were written by workers (except perhaps for St John's Gospel), for workers, and about workers. That is one reason why they have had an influence that no other books have had. There were probably other books written by workers at the same time, preaching more revolutionary doctrines. But they have not survived. Some organized ideology for the Roman workers was a historical necessity. The Roman government even went so far as to invent a special religion for freed slaves, but it did not get very far! In the same way, once the industrial revolution had produced a proletariat, socialist thinkers were bound to arise, and many did. But it was Marx and Engels who laid the theoretical foundations of the only Socialist State.

We must look behind changes in government and ideas to changes in productive forces and relations, and of these, the changes in productive forces are the most fundamental. The first and greatest of these was the origin of production. The earliest fossils that are at all human are associated with tools, if only with crudely chipped stones; and Engels was probably right in saying that an ape-like creature became man when it started making tools. If so this change was the origin of man, not part of his history. Men continued for hundreds of thousands of years as hunters, using fire, stone axes, and later bows and stone-headed arrows. They developed sculpture, painting, and dancing, made huts, and probably wore skin clothes in winter.

Towards the end of the old stone age, dogs joined human society, probably as scavengers. From what we know of the habits of related species, such as jackals, and of the relations of dogs to primitive human societies, it is likely that the initiative came from the dogs, and it is fairly sure that at first they were not private property.

But, at least in Europe, there was a rather sudden change about 10,000 years ago. Men began to domesticate animals and probably to practise a primitive agriculture, and changed over from the use of chipped to polished stone. The domestication of animals or plants as food sources brought about vast economic changes. To take only one of them, the number of people who could live on a square mile of fertile land was increased from ten to a hundred times. This meant that society could no longer be based on a small tribe of twenty to a hundred people who knew one another intimately, and settled most disputes by common sense.

Some sort of government was needed, though not necessarily anything that could be called a state. But domestication meant the possibility of private property not merely in tools, clothes, and huts, but in sources of food, in fact the private ownership of land and capital. Many primitive tribes avoid the development of a class system based on wealth in ways which seem strange to us. For example, in some Pacific islands the produce of the garden worked by a man and his wife goes to his sisters and their children.

Once a man owned sheep or other sources of food, he could and did hire workers without property. The book of Genesis tells the story of how Laban hired Jacob for seven years, and swindled him at the end, while Jacob got the better of Laban at the end of another seven, and became a large-scale sheep-owner. In this kind of way class divisions in society arose. 'The lamb misused breeds public strife,' though not exactly in the way that Blake meant. Class society is quite a recent development, and in no way necessary for human existence or progress.

In fact the men or women who made the great step of domestication lived in a classless society. They must have been people of great intelligence. The idea of keeping a herd of animals with a tribe, instead of hunting it, was a very original one. It may have started in a people with a passion for pets, who began to eat them when they grew too many. It is certainly striking that even in early historical times the Greeks always sacrificed an ox or sheep to a god before eating it themselves, as if they were ashamed of the action, and needed a religious excuse. Even today good Muslims will not eat an animal unless it has been killed in the name of Allah, and many Hindus will not eat animals at all.

Galton believed that only a few animals had instincts which fitted them for domestication, and that all the large domesticable animals had already been tamed. There is much truth in this, as can be seen from the case of the budgerigar, which was only brought over from Australia last century, but is obviously better adapted for human society than any of our British wild birds. However, animal behaviour can be altered by many generations of domestication; and the differences are inherited, as has been shown by crossing wild and tame mice. Today a number of fur-bearing animals such as the silver fox, the mink, and in the Soviet Union the sable, are being domesticated. And their descendants will probably be tame in a few centuries if people continue to wear furs, which is by no means certain, if only because synthetic fibres will probably replace them, as they are replacing silk.

It is very striking that soon after the domestication of animals, art almost disappeared. Neolithic carving is very crude compared with that of the Old Stone Age, and painting is unknown. This may have been due to the origin of classes in society, and a consequent contempt for manual work. A scientific study of what happens when an animal is domesticated will not merely be of interest to biologists, but will help us to understand one of the greatest changes in the past of our own species.

21

Overcrowding at the zoo

When I last went to the London Zoo I was glad to see that the Reptile House had been repainted and that it had a little more light and some more adequate labels stating where the animals came from.

For our zoological gardens are a long way behind those of some other countries.

I think the way in which they were kept before the war was a reflection of imperialism. We had a lot of different animals, and though they were not grossly maltreated they could have been kept a lot better.

The comparison with Scandinavian collections is striking.

We have a number of different species of monkeys in London, almost all in rather small cages.

The Stockholm Zoo has rather few, including only one pair of mandrils. But these animals live in a very large cage and breed quite regularly. There are two babies with the parents and two older youngsters in a separate cage. So one can follow the way in which the male mandril develops his ornamental nose, which is just as interesting as comparing several related species.

Stockholm also has a collection of the larger Swedish-mammals, including brown bears, which live in very large enclosures, and have just produced four charming babies.

At Copenhagen the female hippopotamus regularly produces a baby every second year. She has had nine babies in all. Their giraffes and leopards also breed fairly well.

The plain fact is that in London our animals, like our human beings, are overcrowded. Some day I should like to see a rehousing programme for our Zoo. It could probably be done without using any more land, by putting all animals such as reptiles and tropical birds, which do not live out of doors, on the upper floors of large buildings. Meanwhile, it is obvious that the Zoo cannot be enlarged until our human population has at least as much accomodation as it had before the war. But that is no reason why it should be overcrowded now.

I am not one of those who think that animals are necessarily unhappy in captivity. Here again the experience of the Stockholm Zoo is conclusive. They keep large birds, including herons and eagles in aviaries for some years. Then they let them out, while continuing to feed them regularly. A few fly away; however, the majority do not wander far, but nest in the gardens.

I do not say that this would be possible in London. But it does show that these birds do not mind the conditions in which they are kept. I find it difficult to believe that any bird is happy in a cage so small that it cannot fly about in it.

Apart from humanitarian grounds, there are two reasons why it is important to study the needs of animals, to see what conditions are needed in order that they should live a long time and be fully fertile.

One is this. We know pretty well what food a human being needs for health, and our very successful rationing system is based on this knowledge.

But we only found out what we know by experiments on animals, particularly rats. It had long been known that neither men nor rats did well on a monotonous diet. But no one knew just what was left out in a diet too monotonous for health.

Thanks to experiments on rats and other small animals, we know now. Usually the monotonous diet lacked one or more of the rather varied group of chemicals which are now called vitamins.

If anti-vivisectionists think these experiments were wicked, I can only say that in my experience vast numbers of pet animals, particularly water animals, are worse kept than laboratory rats on deficient diets, and live shorter lives. A tiddler in a jampot may expect a short life, and not a very merry one.

Now if we worked out scientific standards of housing for animals, as we have worked out their food standards, we should discover a number of facts which might be applicable to men and women. Let me give an example.

In an average English home in winter even if there is plenty of fuel, the temperature is very variable. The kitchen is usually hot, and the bedrooms cold. In a centrally heated house the temperature is fairly even. We do not know which of these conditions is healthier for a housewife who spends most of her time indoors. But it is important that we should know before we fix standards for housing.

Animal experiments will not answer the question, but may give us clues to the answer.

The other reason for studying this question is that we have only domesticated quite a few animals and it might be worth trying some more. West Africans, for example, are terribly short of first-class proteins, whether in the form of meat or milk.

Probably several tropical mammals could easily be domesticated, and would suit their needs better than cattle or other species originally derived from more temperate climates. One possible species is the pigmy hippopotamus.

Or, to take a less important example, many children like tame mice. A small, domesticable animal which will live in one's pocket and can be taken to school is clearly a boon. But most mothers object to the smell of mice. Another domesticable species of the same size and with less smell is clearly needed.

For all these reasons I hope that when things become normal again the Zoo will prefer to show us a comparatively small number of animals living under the best possible conditions rather than a large number of slum dwellers.

22

Life slows down for winter

The leaves are beginning to fall, and soon most of our trees will be bare. Annual plants are dead or dying, and so are many animals which only live for a year or less. Others are preparing for winter in different ways. Many birds have already flown south.

Life in fact is beginning to go on at a lower level of activity. We take this sort of thing for granted in Britain. But it is a shock to people from warmer countries. A cousin of mine from New Zealand, on landing in England in winter, thought we must be a decadent old country, as we didn't even trouble to cut down our dead trees.

And it is important to remember that an annual cycle is a special adaptation. Life is in some ways easier for plants and animals where there are no great changes of warmth and light.

The different kinds of plants and animals which live in countries with a cold, frosty winter have adapted themselves to it in different ways.

The winter is difficult for two distinct reasons. Living substance is generally killed by freezing, largely because ice crystals grow in it and break up its structure. And owing to the shortage of light in winter, even evergreen plants cannot make much new substance, so the supply of food for animals is cut down very greatly.

The simplest way of coping with winter is to die in autumn, after leaving a number of eggs or seeds which will not be damaged by frost, and will start the species up again next spring. This is the course taken by many of the smaller plants and animals. It involves special adaptations. In particular, many seeds and eggs will not germinate or hatch till they have been exposed to cold, and then warmed up again. This acts as a safeguard against germination in autumn.

The deciduous trees shed their leaves, and in a great many plants all the parts above ground die off. Before the leaves are shed, most of the living substance is withdrawn from them into the branches and stem. In particular the green chlorophyl is broken down into constituents which can be used again next spring. If the leaves are killed by frost before this process is complete the tree loses a lot of valuable materials. The signal which makes the trees drop their leaves is usually the shortening of the day.

It used to be thought that walnut trees would not grow in Leningrad because the first frosts always came before they had shed their leaves. They can, however, be grown if a tarpaulin is pulled over them daily some hours before sunset for a fortnight or so before the first frosts are expected, so that the length of day is effectively shortened. If so they shed their leaves in time, and make other preparations for winter.

The evergreens protect their leaves from the action of frost in various ways. The most obvious is to fill them with drops of oil and resin which will not freeze. This is the method used by the holly and spruce, and is why holly leaves and spruce needles burn so nicely.

A lot of birds fly to warmer countries in winter. So do a very few butterflies. This means that they can exploit one area in summer and another in winter. No other animals undertake such great annual migrations.

Man could certainly learn from the birds. In a properly organized society every factory and mine-worker who wanted to could get a month's work in the country in summer apart from an annual holiday. This would of course require far better housing in rural areas.

One of the pleasant sides of an agricultural worker's life is that he doesn't do the same work all the year round. In a society planned by workers such variety would be normal.

Most animals eat less in the winter, and put on fat in the autumn which disappears during the cold months. It is used up much less quickly in a cold-blooded animal such as a frog or a snake than in a warm-blooded one such as a field-mouse or a weasel, which has to use food or fat to produce heat during the winter.

However, the cold-blooded animals are liable to be frozen to death if they do not burrow deep into the ground or get into water too deep to freeze.

Some warm-blooded animals, such as the hedgehog, compromise by going into a winter sleep in which their temperature falls, but not all the way to freezing point.

The animals which stay active through the winter mostly eat grass or other animals, and often grow thicker coats.

Tree leaves are mainly eaten by insects, which pass the winter in a dormant condition or as an egg or pupa.

However, a few animals make stores of food. Squirrels and some other rodents store nuts and seeds for the winter. Ants and bees also store food.

Again a fair number of warm-blooded animals protect themselves from the cold by burrowing. They mostly use their burrows in summer too, but clearly a burrow is a very good protection from frost.

The yearly cycle is, of course, not the only one which plants and animals have to face. All except those which live underground or in the deep sea have to cope with the daily round of changes in light and warmth. Some sea animals have daily movements up and down which can be compared with birds' yearly migrations.

Perhaps the most striking rhythmical changes in environment are those produced by the tides. The plants and animals which live in the tidal zone must be able to stand not only the violence of the waves but drying up twice daily, not to mention frost and rain, unless they live in rock pools or burrow into sand or mud.

The most severe conditions of all are those in the estuaries of rivers where the water changes from fresh to salt, and back again, in a few hours. Very few animals can stand up to this. Those few which are adapted to it are often found in immense numbers. Thus aquarists who go down to the Thames in London at low water can get very large numbers of the blood-worm *Tubifex* in a short time; in fact they spend longer in washing the mud off them before giving them to their fish than in collecting them.

Considered as an animal, man has coped with winter by storing food and burrowing from his earliest days. Later on he discovered fire and dressed in the skins of other animals. Still later he made clothes from fibres and built houses above ground.

Storing and burrowing are not production, though they are steps in that direction.

Social insects such as bees and wasps which do not merely store food, but make elaborate combs in which to keep it, may be said to have started production. A hive of bees which can live through the winter because of their productive activities, while most other insects die, is getting rather near the human stage.

In fact, man adapts himself to the coming of winter in many more ways than any animal. But it is worth watching how the animals do it, and even taking hints from some of them.

23

Bird migration

As we looked up to see the Fortresses going over to bomb airfields and factories in France, we saw another section of the cross-channel air traffic, and our song birds and swallows going to warmer countries for the winter.

The main routes of migration are roughly known. The long distance record is held by some of the swallows, which winter in south-west Africa. This is a distance of about 5,000 miles—actually more, as the birds do not fly straight. Of course they alight on the way.

It is not only the birds which migrate. This summer [1943] I have seen a Painted Lady and several Clouded Yellow butterflies. As they never winter over in England, though they can breed here in summer, they or their parents must have flown over from the Mediterranean coasts of Europe, or even from Algeria, where they can live in winter. Only a few of the butterflies of any species in France cross the channel. But there are some species which migrate in masses like birds. The American Monarch or Milkweed butterfly regularly flies north from the southern United States as far as Canada, in the spring, and some at any rate fly south in the autumn. A few members of this species are caught in England; thirty-three is the biggest number in one year, but whether they fly the Atlantic or hitch-hike on ships is not quite certain.

Similarly a few birds migrate in a very irregular way. For example, every twenty-two years or so Pallas' Sandgrouse arrives in England from Siberia in small numbers, and we occasionally get birds from the Arctic or tropics.

We know something about why birds migrate, but nothing about how they find their way. The most important work on the causes has been done by Professor Rowan in Alberta. He showed that even canaries, which are native to warmer climates than ours, can live out of doors in many degrees of frost if they are well fed. But they need a lot of food to keep warm. So he believes that our migrants have to leave because of food shortage rather than cold. What makes them leave is neither cold nor hunger, but the shortening of the day. He kept crows of a migratory Canadian species in a large cage which was floodlit every evening in the autumn, so that the effective length of night did not increase. He found that when released, most of them flew north instead of south, like birds which had had a normal series of nights of increasing length. A number of plants also react to the shortening of the day. Thus many trees only shed their leaves when the days draw in. Walnut trees usually die in Leningrad because the frost nips them before the leaves fall. They will live if covered with a tarpaulin about 3 p.m. early in September, in which case the leaves fall before the first hard frost.

The longer nights act indirectly by making the birds' ovaries and testicles diminish in size and cease to secrete hormones. They do not fly south if the appropriate hormone is injected. And the urge which makes them come back to their breeding places is due to the growth of the same organs in spring. Castrated birds do not migrate regularly. In fact the influence which makes our birds return in spring is the same which later on makes them desire to mate and build nests. We must be very careful in attributing human motives to animals. But the emotion behind migration to breeding places is almost certainly more like human love than hunger or curiosity. The robin is a good example of the exception which proves the rule. It does not usually leave us in winter; and in the autumn its ovaries and testicles increase in size, and produce enough hormones to make its breast redder in winter than in late summer, and, what is more, to keep it at home.

We do not know the answer to the most interesting question, namely how migrants find their way, and particularly how, in some species at any rate, young birds migrate in the right direction without any teaching. This kind of question is commonly called a mystery. I don't like this word. It is taken over from the vocabulary of religion, where it means either something not to be disclosed to the general public, or something which human reason cannot understand. This is just one of the uncounted number of problems awaiting scientific solution. These problems get solved in the long run; for example, we have solved the problems of where the swallows go in winter, and how bees communicate information, both of which baffled our ancestors.

Very likely when we discover the answer it will help airmen to find their way in darkness or fog. At one time it was thought that short wave radio disturbed birds in their flight. I am not disclosing a military secret by remarking that if this were true few birds would find their way to or from England during the war. Nor do magnets put them off their course. Meanwhile we want a lot more information. Before the war thousands of Soviet village schools were doing a co-operative study on bird migration. They caught migrating birds, put rings on their legs, and released them, to be caught again elsewhere. No doubt these children, if they have not been killed or enslaved by the Nazis, are much too busy now. But they or others will start again.

This is one of the problems which is as likely to be solved by ordinary people in their spare time as by laboratory scientists. With the combination of scientific education and leisure to which we may look forward as Leninism spreads over the world, we can look forward to a day when about one person in twenty will be a naturalist, and many mysteries of nature will be mysteries no more.

24

Why the robin sings

Why does a robin sing? The song has been interpreted as courting, and as a sheer expression of joy. Actually both the song and the display of the red breast seems to be a substitute for fighting.

One of the fundamental facts in robin life, as in that of many other birds, is territory. In autumn every male robin occupies a territory of about an acre, in which he spends most of his time though he occasionally feeds outside it. Within this area he sings and struts vigorously. If another male appears in it he sings at him violently, and may peck him. Usually the intruder goes off; occasionally he may fight the quarrel out, or sing it out, and occupy all or part of his rival's territory.

About the new year females arrive in the male's territory, and may or may not stay there. There is a great deal of singing which may serve to attract the females, but the females have an absolutely free choice of mates. One of the visiting females settles down with a male in his territory.

After marriage the female helps the male to defend their joint territory. But they do not mate till March or April, when the nest is at least partly built. At this time the male also brings his wife presents of food, and later helps to feed his children.

Occasionally a female leaves her husband for another male, and two cases of bigamy have been reported, but as a general rule a couple keep together till the young have left the nest. However, during the late summer and autumn the sexes separate, and the hen usually finds a new mate next winter.

The males generally stay in one place for life, but most of the females migrate; some of them passing the winter in France. In fact, the male robin is the home-lover and the female wanders about and chooses her mate.

A robin can live for as long as eleven years. A pair can also hatch out two broods of five or more eggs a year. Now the population of robins is fairly steady. If it increased by only one-tenth per year, it would grow to 13,781 times its original number in a century.

An average pair of robins which live to be adults probably produce at least 20 eggs. And if they lived under very sheltered conditions they could probably produce 100. Actually only two eggs would be needed if there were no untimely deaths.

So we may say that the struggle for life accounts for nine out of ten robins actually begotten, and 49 out of 50 which could be begotten.

A few young robins die in the egg or in the nest, but about three-quarters of those which leave the nest are dead within a year. Even of those which survive the perils of youth, more than half die within the next year. The average life of a robin seems to be about one year or less.

So far as the causes of death are known, cats, mousetraps, and motor vehicles head the list, but probably cold and starvation are equally important.

A robin redbreast in a cage may, as Blake said, put all heaven in a rage; but it is quite willing to breed in it, and is much safer than outside.

25

How bees communicate

EIGHT years ago I gave an account in the *Daily Worker* of the early work of von Frisch and others on the language of bees. In July 1947 I was at the London Zoo with Professor Hadorn of Zurich. We watched bees coming in to the glass-fronted hive laden with pollen of different colours in the bags on their legs. He was able, by watching them, to tell me from what direction they had come, and roughly from what distance. So will you be, after reading this article. The facts previously known were these. When a bee has found a rich source of pollen or honey she comes back to the hive, and before handing it over to the other workers whose job it is to store it in the comb, she does a peculiar 'dance'. During the dance other bees touch her with their antennae, so that they know what smell is associated with the kind of food in question. They then fly off to visit flowers of the same kind, or dishes of sugar-water impregnated with the same smell, for example of peppermint.

When the flower or the sugar-water is placed within 50 yards of the hive, bees fly out in all directions to visit flowers or dishes with the same smell. But when they are placed at distances over about 100 yards, they not merely fly out in the right direction, but for the right distance. As some of them arrive before the original finder has unloaded her honey or pollen it is clear that she must have told them in which direction to go, and how far.

Von Frisch has discovered how the information is conveyed. If the food is within 50 yards the finder always dances round and round. If it is more than 100 the dance is quite different. She goes forward in a certain direction for an inch or two, waggling her abdomen, then runs back without any 'dancing' and repeats the dance again and again. The more she has found, and the sweeter the sugar-water, the longer the dance lasts, and therefore the more other bees are able to learn what smell is associated with food, and the more go to look for it.

If the finder dances the round dance they go out in all directions, but not for further than 100 yards. The other kind of dance gives them the direction. The dances are generally carried out on the comb, but sometimes on the landing stage in front of a hive. If it is horizontal, the dancer moves in the direction of the food, and the other bees fly out in the direction of her dance.

If, however, the surface of the comb is vertical, something much odder happens. As the day goes on, the dancer moves in different directions after coming from the same place. Supposing the food is south-west from the hive, then at 9 a.m. the dancer moves horizontally to the left, at noon she moves at 45 degrees upwards, at 3 p.m. vertically upwards, and so on. In fact a dance upwards means that the food is in the same direction as the sun, a dance to the right that it is to the right of the sun, and so on. It is most remarkable that bees know the direction of the sun, even in cloudy weather. The distance is given by the rhythm of the dance. Food only 150 yards away elicits a dance with 40 tailwags a minute. This number sinks to 20 when the food is half a mile away, and to only 8 at a distance of two miles.

Von Frisch believes that the same language is used by scouts which go out from a swarm of bees and come back to tell it where they have found a place suitable for a new hive. But this is uncertain, for swarming is rare, whereas hundreds of observations can be made every summer day in an ordinary hive.

Besides the dances the bees have at least one other 'word', namely a sweet smell which they make when they have found rich food, and which attracts other bees to the place.

These observations seem to have a great philosophical importance. It is often said that animal 'language' is a mere expression of the emotions, and cannot convey statements of fact. But it is clear that the bees can tell each other not merely that they have found food, but where they have found it. It is true that the bees' language seems to be inborn, and not learned like ours. It is like that of the young lady in Shaw's *Back to Methusaleh* who emerges from an egg talking perfect Shavian English. However, some birds have to learn a good deal of their language. It is also clear that bees have an amazing sense of direction. If a hive is turned round the dancer moves over the comb in a curved path as if she had a compass needle in her head. Perhaps she has some kind of magnetic sense which we lack. Her perception of rhythm must also be superior to our own.

A reader may well ask whether it is not possible that von Frisch is pulling our legs, or at least letting his imagination run away with him, and has invented this amazing story. The answer is that although he made some mistakes in his interpretations of dancing, most of his earlier work has not only been confirmed, but applied in practice by Gubin, Komarov, and others in the Soviet Union, as well as by von Frisch himself in Germany and Austria.

Red clover is normally fertilized by bumble bees, and does not set seed without fertilization. Bumble bees are not common enough to fertilize an area of an acre or more of red clover. And ordinary bees prefer other flowers as their probosces are not long enough to get all the nectar of a red clover flower. The following method is therefore used. Beehives are brought near to the clover fields. Glasses containing sugar-water and red clover blossoms are placed among the clover. Bees soon find them and come back to dance. Their comrades fall for the propaganda and search for flowers with the correct smell. A few of them find the sugar-water. The majority search the clover flowers. They do not find much nectar there, but in their searches they carry pollen from one flower to another. Enough of them find sugar-water to keep up the stream of propaganda in the hive.

The system rather reminds me of the football pools where a few people win large prizes, but the vast majority merely enrich the organizers of the pools and keep the postal workers busy. Economically it pays the seedsmen. For an expenditure of about twelve pounds of sugar per acre over five weeks von Frisch got an increased yield of 36 pounds of clover seed per acre. As a pound of clover seed cost as much as 16 pounds of sugar this was a good bargain, except perhaps for the bees.

Possibly an even bigger return could be got by inducing bees to visit orchards. It is important that the bees visiting prepared sugar-water should if possible perch on fruit blossom and also suck up juice into which blossom has been crushed to give it the right scent. I do not know whether we shall be able to learn the language of ants, and get them to clean our kitchen floors instead of raiding our sugar. But I am quite sure that research on these subjects will tell us things which we need to know, not only about animals, but about human societies.

26

The mysterious eels

Last Saturday a group of the unemployed hooked a number of eels in the trenches which had been dug for Air Raid Protection on Primrose

Hill. It is, of course, a great mystery how these eels got there. But the whole story of eels is mysterious, and it is with other mysteries that I shall deal in this article.

The eel is a fish which has lost its hinder fins, as snakes have lost both pairs of legs and men their tails. But it has a much more surprising deficiency. In all other fish, when full-grown, you can easily find a roe—a hard roe full of eggs in the female, and a soft roe which yields sperm in the male. But the eel is apparently almost sexless. There is a mere rudiment of a gonad or sex gland, and only an expert can tell males from young females.

Thirty years ago the life-history of the eel was only known in fresh water. Where a small stream goes down over rocks directly into the sea, one can sometimes see, from January to April, a swarm of baby eels or elvers swimming up out of the sea. They are generally a couple of inches long and as thick as a stout string. They are extraordinary climbers, and can get up vertical rocks if they are covered with moss.

In large rivers they sometimes form such dense swarms that they have been caught quite literally in tons—an incredible waste of future food. They settle down, preferably in deep and muddy water, and grow from five to twenty years. They are hardy and adventurous creatures, especially the females, and will creep out of the rivers at night through deep grass into ponds. In the days before the water supply of towns was properly filtered, they used to swim up pipes, and were sometimes found blocking taps on the upper floors of houses.

They burrow into the mud and sleep there through the winter, and feed on worms, insect larvae, and small fish during the rest of the year. Females may grow up to 4 feet in length and 25 lb in weight, though males rarely exceed 18 inches. But, however long they are kept in fresh water they never become mature. When fully grown, they swim off downstream to the sea in autumn, and after this nothing was known of their life-history until 1896.

There were many theories. A Scottish naturalist even believed that they were engendered from water-beetles, probably because he mistook parasitic worms in these beetles for young eels. The problem was solved, at least in part, by the great Danish zoologist, Schmidt.

If you drag a very fine net through the North Atlantic you sometimes catch a small transparent flat fish about the size and shape of a willow leaf or smaller. In 1896 the Italian, Grassi, had discovered that these fish never grew beyond 3 inches long and then change into young eels, so tadpoles change into frogs, losing three-quarters of their weight in doing so.

As Schmidt sailed about the Atlantic he found that the transparent eel larvae got smaller as he approached an area beginning between Bermuda and the West Indies, and extending east for about 600 miles. Here the sea is over four miles deep. The eels from an area ranging from Iceland to the Canary Islands and from the Azores to Cyprus swim across the ocean to this abyss, becoming sexually mature as they go. Here they mate and die. This behaviour is found not only in heroes and heroines of Wagner's operas, but in other migratory fish. European salmon have the opposite habit to the eel. They do most of their growth in the sea, and swim up the rivers to mate, after which some of them die, but a great many swim down to the sea again, and come back next year. But the salmon of the Canadian Pacific coast, which we buy in tins, invariably die after mating.

How do the little eels find their way across the Atlantic? The answer probably is that they don't. They are carried by the Gulf Stream and are three years old when they reach Europe. Schmidt argued as follows. Eels are found in Western Europe from the Adriatic to Scandinavia. If the eels derived from British parents came back to Britain and those of Italian origin to Italy, there would be different local races of eel.

Now the number of vertebrae in the backbones of many fish vary. And Schmidt showed that in trout and other fish the average number varied in different local races, and was partly hereditary. But in eels from all parts of Europe the average was exactly the same, somewhere between 114 and 115.

On the other hand, the fresh-water eels of the West Indies and those parts of North America which drain into the Atlantic have a breeding ground close to that of the European eels, and have a different number of vertebrae, usually 106. So there are two quite different races or species of fresh-water eel, besides, of course, salt-water eels, such as the conger.

Near the Bermuda Islands the young of both sorts of eel are found together. They are sorted out, not by homing instinct, but in a simpler way. The American race become elvers, with a desire for fresh water, when they are only one year old and the swarm of little fish is still near the American coast. The Europeans do not change until they are three years old, by which time they are mostly near Europe.

How the adult eels find their way back to their place of birth and death is quite unknown. Possibly by memory, and perhaps by means of some sense which men do not possess, though many birds possibly do. Besides this problem, there are plenty more to answer. Some day a ship equipped with a special trawl to work at a depth of five miles will go to the eels' breeding ground, and we shall find out what an adult eel looks like. Later on we shall discover whether cold salt water is enough to make an eel mature, or whether it needs high pressure as well. But the eel is a slippery customer in more ways then one, and will probably keep biologists busy for many years to come.

27

Movies for toads

PRIMITIVE men take it for granted that animals can think, and according to many religions they have souls which are judged after their deaths. Christian philosophers have usually denied that any other animals were capable of reasoning or had any rights. So, till the nineteenth century at any rate, animals were better treated in India than in Europe. Darwin took the view that the higher animals possessed most of the human faculties, though many of them poorly developed. On the other hand some of the more mechanistic biologists try to explain all animal behaviour on mechanical lines.

It is extraordinarily difficult to be sure that animals are thinking, and not picking up clues given by the experimenter. For example about 1910 a German at Elberfeld had some horses which could do fairly elaborate arithmetic, such as extracting cube roots. A French journalist, zealous for his country's honour, produced the story of a cat at Bordeaux which corrected the children's homework, mewing when there was a mistake in a sum and purring when it was right. However no one ever saw this cat, and plenty of people saw the horses. When a sum was put up on the blackboard, they tapped out the answer with their hooves. But they did not do it unless the teacher was there, and the psychologists who examined them finally concluded that they watched him carefully, and stopped tapping when he wished them to. Perhaps he made some signal deliberately; more probably they noticed slight changes in his expression or breathing.

A better idea of an animal's capacity for grasping numbers comes from the experiments of the late Dr Honigmann on hens. He put hens in a cage with a narrow gap in the floor. Under this gap a board moved on rollers, carrying a row of wheat grains, of which the hen could pick one at a time. He glued down every second wheat grain, so that the hen could not remove it. After a while some hens learned only to pick at alternate grains. But they failed completely to conform to the situation when only every third wheat grain was free. If we like to put it that way, we can say that a hen can count up to two, but not to three. Other birds seem to be able to count up to five or so, or at least to notice a difference between four and five eggs.

Another of Dr Honigmann's experiments was probably still more interesting. It has often been stated that animals cannot appreciate pictures. Certainly they recognize gramophone records. And female crickets will come to a telephone if a male of their species is chirping at the other end. But a dog rarely, if ever, shows any interest in a picture or photograph of his master or another dog. Nor is he interested in cinema films.*

The first animals which undoubtedly reacted to a moving picture are the common toad and the natterjack toad. They have very simple minds, if mind is the right word. They will only eat moving objects. The response to food is very characteristic. Although it is slow in its ordinary movements, a toad can flick out its tongue with very great speed and accuracy, and bring back a small insect or other food into its open mouth. It then swallows it if the taste is suitable. They are not interested in dead mealworms if they stay still. But a toad will flick and snap at a dead mealworm if it is dragged past the toad, especially if there is a well-marked background behind it. The toad snaps in the same way if the mealworm stays still, and the toad and the background are dragged past it. Toads will also snap at a film of a moving mealworm or other small animal. It might be argued that they react mechanically to any small object which moves or seems to be moving. But this is not the whole story.

If two toads are competing for the same food, and one of them gets it, the disappointed toad, especially if hungry, may flick its tongue at the eye of its successful rival. This is particularly common if the successful toad is the smaller of the two. Honigmann made films of toads eating, and showed them to other hungry toads. I have myself seen them flicking the film star in the eye with their tongues. But this is most constantly done if the film is shown a little less than life size, and rarely happens if it is enlarged. The toad is not a noble animal. A toad will attack a picture of itself feeding as readily as that of another toad, but it is quite uninterested in films of other small animals, such as hamsters, feeding.

Encouraged by his success with toads, Honigmann went on to show his films of moving worms to dragonfly larvae. These live under

* Several correspondents claim that their dogs enjoy films, particularly of the 'Wild West' type.

water, and shoot out their jaws at their prey. The film had to be shown projected onto a sheet of white paper pasted on the side of a glass tank, and the insects snapped at the pictures.

It is curious that so far moving pictures have only interested animals with very simple minds, such as toads and dragonfly larvae, and those with very complicated ones, namely ourselves. I think the reason is clear. A dog or a cat has mind enough to know that the moving picture is not a real happening, though a few dogs will watch films representing vigorous action. Men alone have negated this negation. We know that the hero is not really in danger of the electric chair, and that we shall not meet the heroine even if we stay at the stage door all night. But in spite of this, we manage to work up the appropriate emotions in a minor key, so to say.

I have described these particular experiments rather than hundreds of others which are constantly being made on animal behaviour, largely because I liked Dr Honigmann, a refugee who worked at the London Zoo before the war, was interned, and carried out his work on toads in Glasgow until his death. I liked him partly because he so obviously liked toads. While one can go too far in treating animals like human beings, I believe that one achieves the best results, whether with animals, plants, or men, if one likes them as well as being interested in them.

28

The chicken or the egg?

The most frequently asked question is: 'Which came first, the chicken or the egg?' The fact that it is still asked proves either that many people have never been taught the theory of evolution or that they don't believe it.

For the answer is quite clearly 'The egg'.

If by 'chicken' you mean the young of our domestic poultry, poultry were descended from other birds which laid eggs, and at whatever point you choose to draw the line, the first chicken came out of an egg, and this egg came out of something which was not quite a chicken.

If by 'chicken' you mean any young bird, then there is no serious doubt that birds are descended from egg-laying reptiles, and the answer is therefore the same.

This is so even though, if we had specimens of all the ancestors of a living bird species, we probably should not know where to draw the line. The earliest, bird-like animals whose fossils are known had teeth, long, bony tails, and a line of flight feathers between their front and hind limbs.

They probably glided like the Australian 'flying' phalanger, and it is doubtful if they could in the ordinary sense of the word. So they may not deserve the name of birds.

Some reader may ask why I am so certain. The answer is that a scientific theory becomes certain in proportion to the number of times when an experiment or an observation might have disproved it, and has not done so.

Now the theory of evolution consists of two parts. One part is the theory of descent, namely, that existing animals are descended from ancestors unlike themselves, but that the changes in a single generation have never been very great. The second part of the theory deals with the causes of evolution; for example, Darwin thought (and I agree with him) that natural selection was the main cause.

A great many lines of evolution are well established by the study of fossils.

We have a very good genealogy of the horse, a fair one of man, and a very poor one of birds. Few birds get fossilized, as the acid soil produced by decaying leaves in a forest destroys bones; and most birds live, and probably lived, in forests.

However, the broad lines of descent are so clear that evolution would at once be disproved if the skeleton of a man, a horse, or a bird were discovered imbedded in a coal seam.

In case any practical joker wishes to claim that he has found one there, I may add that bones from the coal measures are impregnated with vegetable material, and a fake would at once be detected.

On the other hand it is not so certain how and why evolution occurred. A few able students of evolution think that natural selection was not the main force behind it, and adopt Lamarck's explanation or some other. Many more biologists suspend judgement.

The situation in human history is much the same.

It is quite certain that the English annexed the Gold Coast. Marxists think the reasons for this were economic, but they are prepared to argue with people who say the motive was a desire for glory, or a wish to spread civilization or Christianity in Africa.

It is no use arguing with a person who says that the Gold Coast is independent, or belongs to Brazil.

I have chosen a very simple example of a question which appeared insoluble before we had our present knowledge of evolution. Further advances in science will answer other questions which seem insoluble at present.

For example, if the views on the evolution of the stars which Milne and I are propounding at present turn out to be right, a lot of ancient difficulties about time and space will vanish.

But this theory will have to be checked, as the theory of evolution has been checked by a study of fossils. And the checking will take decades or, more probably, centuries.

The detailed consequences will have to be worked out, and astronomers will have to look for facts which would disprove the theory if it is wrong.

Engels, and also some non-socialist thinkers, such as W. K. Clifford, saw that evolution cleared up many more important problems than that of the egg and the chicken. For example, people still speak of eternal values, or eternal laws of morality.

There is nothing eternal about the prohibition of stealing. Less than a million years ago our ancestors had nothing which could be called property. They did not store food, use tools, or wear clothes.

In the remote future there may be no private property either because all useful things will be made in such quantities that we shall no more want them for private use than we now want air or sunlight, or because the feeling of human brotherhood will be so strong that we shall take common ownership of everything as a matter of course.

Again, evolution makes idealism much more difficult. If I were the only sentient being in the world there would be no way of deciding whether things were really there, or only in my mind. I am not sure if there would be any sense in the question.

The fact that many people can see and feel the same mountain makes it much harder to believe that the mountain is nothing but an idea, a permanent possibility of sensation, or something of that kind.

But the theory of evolution and the geological record make it reasonably certain that there were mountains long before there were any animals, let alone human beings, to be aware of them, and Engels regarded this as one of the strongest arguments against idealism.

In fact a belief in evolution should influence our thought on almost, if not quite all, subjects.

Its importance for communists can be judged from the fact that Marx wanted to dedicate *Capital* to Darwin.

29

The mathematics of evolution

The greatest difficulty in explaining science to ordinary people is that almost every part of it is becoming mathematical. The mathematics are not always very difficult. For example you do not need much more mathematics to study heredity than to study contract bridge. But you do need some.

One of the studies which is rapidly becoming mathematical is that of evolution. Darwin thought in words. His successors today have to think in numbers. Everyone who has gone into the evidence, which takes some years to do, believes in evolution. That is to say he believes that the animals and plants living today are descended from very different ones in the past, some of which have left fossils. But there is a good deal of doubt as to some of the lines of descent and an immense amount about how evolution happened.

Most biologists think an explanation based on natural selection will account for it. But some believe, with Lamarck, that acquired characters are inherited, for example that if you feed a hen well, not only will it lay more eggs, but so will its daughters. Others believe evolution is divinely guided, in spite of the fact that this puts the responsibility for the tapeworm and the tubercle bacillus on God (for there were certainly parasites long before there were any men to sin). Still others say they don't know.

The first place where mathematics come in is in fixing the time scale. This can be done by analysing radioactive minerals. For uranium and thorium gradually transform themselves into lead, which has a different atomic weight from ordinary lead, and the older a radioactive mineral the more of this special type of lead it will contain.

The next step is to measure a number of fossils carefully in order to see just how much change has occurred in, say, two million years of evolution. The results are astonishing. The teeth of horses have been getting longer for some 50 million years. Their ancestors were browsers, that is to say they ate the leaves of trees, for which they only needed short teeth. But grass is a good deal grittier than tree leaves, besides containing grit from the soil, and wears the teeth down. So a short-toothed animal could only live for a year or so on grass. It would die when its teeth were worn away. But teeth have changed so slowly that if you measure corresponding teeth from a population of fossil horses and from their descendants two million years later, although the average values have changed, there is often still some overlap. That is to say the shortest teeth two million years later are no longer than the longest two million years past.

The next step is to see if you can change the characters of a population by exposing it to natural selection under controlled conditions. This has been done with populations of flies by Dubinin in the Soviet Union, by Dobzhansky in the United States, by Kalmus in England, and above all by Teissier in France. The mathematical theory of these changes is fairly complicated, and a part of it was worked out by myself before any of these experiments had been done, while Wright and Fisher have tackled some of the still more complicated problems which arise in natural evolution.

Curiously enough we know more about natural selection in man than in any other animal or plant. The reason is a simple one. One can study human beings with various inherited abnormalities and see how long on average they live and how many children they have. One cannot do this with wild animals. White mice in captivity are just as fit as coloured ones. They live about as long and have as many children. And whites do not disappear from a mixed population. But they are less fit in the wild state, probably because they do not see as well as normal mice, and are more conspicuous to their enemies. However, one cannot study a thousand wild white mice and a thousand coloured ones, and see just how the white ones are less fit. One can make such studies on hundreds of human dwarfs or haemophiliacs, that is to say boys whose blood clots very slowly.

So the most immediate application of the mathematical theory of natural selection has been to human society. Unfortunately most of the ladies and gentlemen who wish to improve the human race seem to find the theory a bit too stiff. I do not blame them for finding it stiff. I do blame them for putting forward eugenical schemes without the necessary mathematics. This is as futile as trying to design a high-speed aeroplane without mathematics, and a lot more dangerous. For a badly designed aeroplane will probably only kill a few pilots and passengers. But false ideas about racial biology may lead to the death of millions, as Hitler demonstrated.

In fact the theory shows that some 'racial hygiene' is possible, but that it is far less efficient than has been thought. We could prevent about half the haemophiliacs from being born, and about a quarter of the dwarfs. In either case we should have to interfere with human liberty to some extent. I doubt if it would be worth while. It would be still harder to stop mental defectives from being born, for the good reason that most of their parents are normal. This does not mean that it will always be impossible either to prevent the birth of such children, or to treat them so that they grow up into rational people.

To come back to evolution, I think it has been proved that natural selection is an effective agent, and will explain a very great deal of what has happened. But some changes are certainly harder to explain than others; and I think it is still an open question whether all evolutionary change can be explained in this way. What I am sure of is that it is as useless to argue about some of these doubtful cases in words as to argue in words about whether or not an aeroplane will ever fly faster than sound.

The fact that science is getting more mathematical is one of my main difficulties in explaining it. The remedy is for children to learn more mathematics, which they could do if mathematics were brought into relation with real life, instead of with ridiculous problems about the price of eggs, which is controlled anyway. But till my readers know more mathematics, I have to write more dogmatically than I like.

30

Back to the water

Much of my work during this war, some of which I have been allowed to describe, has been in connexion with human life under water. Naturally I had to see if I could get any hints from other air-breathing animals which have taken to life in water, and about whose physiology something is known.

Both the study of fossils and that of comparative anatomy make it fairly clear that there were animals in water before they came on land. And they leave no doubt that the four-footed land animals are descended from fish which came out of the water about the time when the old red sandstone was laid down, before the coal was formed. Probably we and vertebrates are all descended from a single species of adventurous fish. But very many groups of land vertebrates have gone back to the water. Some, like water-voles, otters, gulls, and sea snakes, show no very obvious changes in their anatomy. But these spend a good deal of their life ashore, and above all their young are born or hatched in the air.

Other aquatic animals such as seals, penguins, and turtles, are so far transformed that their limbs are greatly modified for swimming, and neither seals nor turtles can walk far, while penguins cannot fly. But they still come ashore to bear their young or lay their eggs.

Only two groups of mammals have been fully modified for aquatic life, and live their whole lives in the water. These are the flesh-eating whales and dolphins, and the vegetarian sirenians, such as the manatee. No bird has ever managed this, and the ichthyosaurs, an extinct group of reptiles which did so, brought forth their young alive, which is fairly unusual in reptiles, but obviously necessary in an air-breathing marine animal.

The same kind of thing has happened in the evolution of insects. Their remote ancestors were aquatic, and many different groups have gone back to the water. Again this is usually only for part of their life cycle. Thus dragonflies, mayflies, caddis flies, and mosquitoes spend their larval stage under water, and do all their growth there. They only come out at their last moult, mate in the air, and lay their eggs on or near water.

In the course of evolution there are comparatively few examples of water animals adapting themselves to land, and a great many of land animals adapting themselves to water. And it is a striking fact that many of the land animals which have gone back are more efficient in the water than its original inhabitants. The largest whales are larger than any fish of the present or past. And they almost certainly swim quicker, grow faster, and are more intelligent. As they keep their temperatures steady, they can live in warm or cold water, and thus have a wider range north and south than any fish.

Clearly life on land has given them some useful characters which they have taken back to the water. A fish coming out of water finds itself in a more difficult environment than before. It is not supported on every side. It must develop limbs, which need much more complicated brain function to work them than fins. It needs stronger bones, and must give its young a tough egg shell or bring them forth alive. It is subject to a greater range of temperatures, and it is an advantage to it to be able to keep warm in cold weather and cool in hot weather. A constant temperature is probably needed for the high development of brain function found in mammals and birds. A 'cold-blooded' animal, more accurately an animal of variable temperature, probably could not develop great mental powers. Certainly the human brain is more upset by a rise or fall of temperature than our other organs.

So the adaptations developed by land animals to meet the difficulties of life on land, of which I have only mentioned a very few, have proved useful to their descendants which went back to water, even though they continue to be air breathers, and must come up to the surface from time to time. Seals, porpoises, and probably whales, manage to dive for many minutes without suffocating, by cutting off the blood supply from most organs except their brains, and slowing down their hearts so as to conserve oxygen. Human divers cannot do this, so if they are to stay down even for five minutes they must have air pumped down to them, or take compressed air with them in cylinders.

The facts of animal evolution have a considerable bearing on the development of human societies. Of course, however, one must be careful not to argue uncritically from one to another. But the history of whales and the like may help us to understand why scientific communism could not develop directly from primitive communism, but a period of class society was inevitable. This is important if we are to see through the arguments of some anarchists, simple-lifers, and others who want us to discard many of the good things of capitalist civilization along with its evils.

A member of a primitive communist society is bound by custom, and seldom seems to think for himself, as Engels pointed out. Worse still, he may treat the members of his own tribe like brothers and sisters, but find his greatest pride in collecting the heads of a tribe five miles away. Apparently class society was necessary to develop the division of labour, to allow the formation of communities so large that each member does not know every other, and above all to develop technology beyond that of the stone age. In class society each man must fend for himself, and thus develop intellect, if not morality.

We now know that it is possible to keep these gains while abolishing the class distinctions which helped to generate them. There is no more reason to suppose that men in a scientific communist society will go back to the primitive mental and moral processes of primitive men, than that whales will become coldblooded or give up suckling their young.

31

The common cold

This is the worst season of the year for common colds. From one to three or more times a winter, most of us are inefficient for several days as a result of this infection. In normal times the most public-spirited thing to do if you had a cold was to knock off work, so as not to affect your mates. At present I think we are supposed to stay at work and trap the germs in our handkerchiefs. The *Lancet* recommends us to wear a pad over our noses and mouths for three days, but it is not so easy to get the material.

There is no doubt that a 'cold' is due to infection. This has been shown in a great many ways. Careful studies have been made in small islands where no one has had a cold for many months until a ship arrived, and then it has gone round the whole population. The agent is too small to be seen even with a microscope, for colds have been given, both to men and apes, by filtrates of nasal secretions which had passed through a filter so fine as to stop all bacteria. Apes get our colds, and get them very badly. The glass screens between us and them in the Zoo are to protect them from our airborne diseases, including tuberculosis, but particularly colds.

While a virus is onc cause of colds, it is not the only cause. The weather is somehow concerned, which is why colds almost disappear in summer. But it is not the main cause. Arctic explorers never get colds, until they come back to 'civilization' and get real bad ones.

Nobody knows where the virus of the common cold spends the summer. Even in warm weather there may be enough people with colds to keep it going. Perhaps a few people can carry it without showing any symptoms. Until this is known there is no prospect of wiping out colds.

Attempts to prevent them with vaccines, serum, and so on, have been a failure, or at least have not been very successful. An English professor injected half his medical students with a vaccine which was supposed to be prophylactic. He asked them if it had done them good, and the majority said yes. But he also made them keep diaries of their colds, and found that they got just as many, and as bad ones, as the untreated half. Other workers have of course claimed better results.

Colds can probably be cured with some of the new drugs related to sulphanilamide. But these are dangerous substances, and can cause illness far worse than a cold. The risk is worth taking in the case of blood poisoning, pneumonia, or gonorrhoea; it is emphatically not worth taking for a cold.

However, a very great deal can be done for a cold with ephedrine. This drug is derived from a root which has been used in China for a long time under the name of Ma Huang, *Ephedra* being the scientific name for the plant. Many traditional Chinese medicines are quite worthless, as are most of the traditional European ones, including a good many which are still prescribed by doctors and sold by manufacturers. Ma Huang is rather uncertain in its action, as the amount of ephedrine in different roots varies. This is generally so with herbal remedies.

Chinese and Japanese scientists have investigated a number of these medicines, and the most valuable substance so far found in them is ephedrine. Its chemical formula is similar to that of adrenaline, the substance which the adrenal glands, lying close to our kidneys, pour into the blood during violent emotion or exercise. Adrenaline, among other things, contracts the small arteries and speeds up the heart. But its effects do not last very long, or our hearts would go on thumping for hours after we had keyed ourselves up when we run to catch a train or hear the sirens announce an alert. We have chemical means of destroying it in a few minutes, and making more when it is needed.

Ephedrine has most of the effects of adrenaline, but we do not destroy it rapidly, so they last for some hours. It can be used locally, as surgeons use adrenaline to stop bleeding from small blood vessels. A number of solutions of ephedrine for dropping into the nose are on the market, and some of them are pretty effective. If you treat your nose with one of them before going to bed the blood vessels in the inflamed membrane shut up, and I, for one, can breathe through my nose again. This not only enables me to go to sleep. It helps me to sleep with my mouth shut so that I do not breathe cold air, and the infection is much less likely to spread to my throat and lungs. It also stops me snoring.

One can also eat ephedrine. I do so when I have a cold, and it certainly stops my nose running. But it puts my pulse rate up to 100 or so per minute, and raises my blood pressure. As my blood pressure is normally very low, I don't mind; but no one should eat it unless they are sure that their blood pressure is normal or low, and I don't recommend anyone to start on more than half a grain. Ephedrine is also liable to keep you awake if you take it by the mouth. But I certainly recommend it. I made two speeches on January 30th and 31st, and I doubt if any of my audiences spotted that I had a fairly bad cold—or should have had but for ephedrine.

Very likely some other compound will be made which, for the same action on the nose, has less effect on the heart than ephedrine, and less on the brain than benzedrine, which is also of some value against colds. Obviously, systematic work on these lines will not be done during the war.

And even in peace the medical profession is not much concerned with colds. They are not consulted about them, and as we only pay them when we get ill, they can hardly be expected to do much research on the subject. Yet colds cause a vast amount of unhappiness and inefficiency. When the economic basis of the medical profession is changed so that their main function becomes, not merely to cure or prevent serious disease, but to keep us in perfect health, colds and many minor illnesses will be attacked as vigorously as typhoid or smallpox. But this is hardly likely to happen until medicine is socialized.

32

Medical measurements

In another article, I wrote about the physiology involved in the *Thetis* case. My evidence depended on thinking quantitatively about how human beings work. It is no good just saying that carbon dioxide is poisonous. You could breathe air containing 2 per cent of it for a lifetime. But if you breathe air containing 6 per cent, you pant violently and become rather ill. A little of it is harmless. Too much of it is a poison. We have a case of the change of quantity into quality.

We are only just beginning to make accurate measurements of living beings. The most familiar of these are, of course, the human height and weight. Men have been measuring and weighing for several thousands of years—in fact, since cloth, grain, and metal became commodities. For a commodity must be accurately weighed or measured.

Another familiar measurement is the pulse rate. But this could only be measured when watches or clocks with second hands had been invented. And they were not invented till they were needed. The monks who made the first clocks did not mind if they were a minute late or early in starting Compline. Accurate clocks were first made for navigation, so that by looking at the sun or a star and a clock at the same time, sailors could tell how far east or west of Greenwich they were.

A rapid pulse is a sign of fever, but it is much better to measure the temperature. This familiar measurement is based on a very recent invention. Sir Clifford Allbutt, who introduced clinical thermometers into Britain, only died fairly recently. And they are now so cheap and familiar that we forget that they involve thousands of years of work, and that a few hundred years ago people did not distinguish between the 'heat' of mustard and that of boiling water.

Another familiar medical measurement is that of the blood pressure, or, more accurately, of the maximum pressure in our arteries. The doctor puts a rubber sleeve round your arm, and raises the pressure in it till he can no longer feel or hear the pulse in your wrist. If the pressure is too high, your life is likely to be short, and you are particularly liable to a stroke of paralysis from a burst blood vessel in your brain.

Other measurements are chemical. The composition of your blood is kept extremely steady, because the various cells in a human body, though much more efficient at their jobs than those of simpler animals, can only work if they have the right environment. So blood analysis is becoming important.

Scores of different blood constituents can be measured. But each method for measurement is the outcome of hundreds of years of laboratory work of which very little is heard outside scientific circles. If we want to measure, say, the amount of uric acid in blood, we must first add a chemical which will precipitate a number of other blood constituents, then filter off the precipitate, leaving a clear fluid, add another chemical which produces a blue colour with uric acid, and, finally, compare the depth of colour with that produced by a standard amount of uric acid in a special instrument.

And this is one of the easier blood constituents to measure. However, I think it would take most of my readers several weeks to learn to measure it with reasonable accuracy. One of the first things you must learn is how to wash glass so that it is clean from a chemist's standpoint, which is much more exacting than a housewife's. I well remember the first blood analysis I made in a hospital. If some constituents were sufficiently near normal a man was to be operated on. Otherwise it would be too risky. If my analysis went wrong, I was condemning him to death.

As each new method is devised and finally made accurate, it is first necessary to determine the normal value of what is measured, and its range of variation in healthy people. Then it is studied in people with various diseases. A low blood phosphate is a sign of rickets, a high blood phosphate of kidney disease, and so on.

In fact, living creatures turn out to have a chemical structure just as definite as their physical structure. There are chemical freaks, which may or may not be invalids. Sometimes we breed them, if the chemical change gives us a new colour in a flower or an excess of butter in milk.

But all this knowledge is still very much up in the air. In a few great hospitals and clinics, the new methods are being fairly thoroughly applied. But though the productive methods in medicine have changed utterly, the productive relations, except for a few details, are what they were centuries ago. We pay the doctor if we are ill, and hope he will be honest enough to cure us as quickly as possible and cut his own fees.

I want to see Socialist medicine, but it is worth noting that medicine is not yet even in the capitalist stage of development. You go to an individual doctor, as your great-great-grandfather went to an individual weaver. The doctors are bound by a professional code like that of a Mediaeval guild, which ensures a fairly high standard of work, but slows down progress.

The socialization of medicine will mean an even greater revolution in the doctor's work than in that of the factory hand. And until medicine is socialized, many of the methods of which I am speaking will remain scientific curiosities of little value to the ordinary man or woman.

33

The differential calculus

No elementary school child gets a chance of learning the differential calculus, and very few secondary school children do so. Yet I know from my own experience that children of twelve can learn it. As it is a mathematical tool used in most branches of science, this forms a bar between the workers and many kinds of scientific knowledge. I have no intention of teaching the calculus, but it is quite easy to explain what it is about, particularly to skilled workers. For a very large number of skilled workers use it in practice without knowing that they are doing so.

The differential calculus is concerned with rates of change. In practical life we constantly come across pairs of quantities which are related, so that after both have been measured, when we know one, we know the other. Thus if we know the distance along a road from a fixed point we can find the height above sea level from a map with contours. If we know the time of day we can determine the air temperature on any particular day from a record of a thermometer made on that day. In such cases we often want to know the rate of change of one relative to the other.

If x and y are the two quantities, then the rate of change of y relative to x is written $\frac{dy}{dx}$. For example, if x is the distance of a point on a railway from London, measured in feet, and y the height above sea level, $\frac{dy}{dx}$ is the gradient of the railway. If the height y increases by 1 foot while the distance x increases by 172 feet, the average value of $\frac{dy}{dx}$ is $\frac{1}{172}$. We say that the gradient is 1 in 172. If x is the time measured in hours and fractions of an hour, and y the number of miles gone, then $\frac{dy}{dx}$ is the speed in miles per hour. Of course the rate of change may be zero, as on a level road, and negative when the height is diminishing as the distance x increases.

To take two more examples, if x is the temperature, and y the length of a metal bar, $\frac{dy}{dx} \div y$ is the coefficient of expansion, that is to say the proportionate increase in length per degree. And if x is the price of a commodity, and y the amount bought per day, then $\frac{-dy}{dx}$ is called the elasticity of demand.

For example people must buy bread, but can cut down on jam, so the demand for jam is more elastic than that for bread. This notion of elasticity is very important in the academic economics taught in our universities. Professors say that Marxism is out of date because Marx did not calculate such things. This would be a serious criticism if the economic 'laws' of 1900 were eternal truths. Of course Marx saw that they were nothing of the kind, and 'elasticity of demand' is out of date in England today for the very good reason that most commodities are controlled or rationed.

The mathematical part of the calculus is the art of calculating $\frac{dy}{dx}$ if y has some mathematical relation to x, for example is equal to its square or logarithm. The rules have to be learned like those for the area and volume of geometrical figures, and have the same sort of value. No area is absolutely square, and no volume is absolutely cylindrical. But there are things in real life like enough to squares and cylinders to make the rules about them worth learning. So with the calculus. It is not exactly true that the speed of a falling body is proportional to the time it has been falling. But this is close enough to the truth for many purposes.

The differential calculus goes a lot further. Think of a bus going up a hill which gradually gets steeper. If x is the horizontal distance, and y the height, this means that the slope $\frac{dy}{dx}$ is increasing. The rate of change of $\frac{dy}{dx}$ with regard to y is written $\frac{d^2y}{dx^2}$. In this case it gives a measure of the curvature of the road surface. In the same way if x is time and y distance, $\frac{d^2y}{dx^2}$ is the rate of change of speed with time, or acceleration. This is a quantity which good drivers can estimate pretty well, though they do not know they are using the basic ideas of the differential calculus.

If one quantity depends on several others, the differential calculus shows us how to measure this dependence. Thus the pressure of a gas varies with the temperature and the volume. Both temperature and volume vary during the stroke of a cylinder of a steam or petrol engine, and the calculus is needed for an accurate theory of their action.

Finally, the calculus is a fascinating study for its own sake. In February 1917 I was one of a row of wounded officers lying on stretchers on a steamer going down the river Tigris in Mesopotamia. I was reading a mathematical book on vectors, the man next me was reading one on the calculus. As antidotes to pain we preferred them to novels. Some parts of mathematics are beautiful, like good verse or painting. The calculus is beautiful, but not because it is a product of 'pure thought'. It is not a product of pure thought. It was invented as a tool to help men to calculate the movements of stars and cannon balls. It has the beauty of a really efficient machine.

To judge from the technical books which sell by tens of thousands in the Soviet Union, a bigger fraction of the people understand it there than here. In a society where workers are encouraged to understand their work it is natural that it should be widely studied. Those who are working to build such a society in our own country, even if they cannot yet learn it, should know a little of what it means.

34

What is the fourth dimension?

When we say space has three dimensions we mean that a position in space is determined by three measurements: for example, so many miles north, east, and up from a definite place, such as the *Daily Worker* office.

If we like, two of the measurements may be angles; for example, 37 degrees east of north, and 5 degrees above the horizontal, but one must be a distance or involve a distance.

Three measurements are always enough, and a fourth either tells you nothing more or makes nonsense. For example if we say that an aeroplane is 12 miles north of our rotary press, four miles east, and three miles up, we say nothing fresh by adding that it is 13 miles away, and we talk nonsense if we say it is 15 miles away.

It is a hard fact that space has three and only three dimensions, as it is a hard fact that I have two and only two eyes. Nevertheless, mathematicians lecture and write about the geometry of four, five, or many-dimensional manifolds or spaces. I even refer to it myself in one course of lectures.

The main justification is this. We use ordinary space for representing mathematical relations by means of graphs. For example, you can follow the changes in employment or real wages, or the population of Britain, much more easily from a graph than from a table of figures.

Clearly, if we are merely considering the relation between two numbers we only need two dimensions, and can make our graph on a sheet of paper. For example, if we want to show the frequencies of different heights in the population, we could draw a graph showing that 20 per cent of the men in a group had heights between 69 and 70 inches, 12 per cent between 68 and 69, 10 per cent between 67 and 68, and so on falling off on each side of the maximum.

If we want to do the same thing for three numbers we need a solid model. For example, we can put up a wire on each square of a piece of squared paper to represent the frequencies of married couples in which the husband measures 69–70 inches and the wife 62–64, and go on. We can then make a surface through the tops of these wires, and get a solid model. When we look at it, we see at once there is a definite, but not very strong, tendency for tall men to marry tall women, and so on. There are plenty of exceptions, but no doubt about the tendency.

Now in science we constantly want to do calculations involving relations between four numbers, for example, the frequencies of different height combinations of father, mother and child. We can't make a graph or model of such a relation, but we can argue as if we had made one. We can say that our four-dimensional model would or would not be symmetrical, that it would have one or several 'peaks' or maxima, and so on.

And just as various sections of a three dimensional model are two dimensional graphs, for example contour lines, so various 'sections' of the four-dimensional imaginary model can be represented by solids.

Again, if we have two variable numbers, say, the pressure of steam in a cylinder and the position of the piston, it is most useful to the engineer to graph one against the other. In the course of a stroke the point moves round a certain area, which measures the work done.

We often have to deal with such relations between several numbers, and we need to be able to calculate quantities corresponding to volumes, solid angles, and so on, in many dimensions.

In fact the geometry of many dimensions is of practical importance. It is also quite amusing.

No doubt all the practical results can be reached by other methods. But it is worth noting that the pioneer in using many-dimensional 'models' in mechanics was Hertz. And Hertz was also the first man to produce and pick up radio waves; so it is no good dismissing him as an unpractical theorist.

If we want to represent the intensity and direction of gravity, and of electric and magnetic forces at any point in space, we can do so by using extra dimensions. There is nothing mysterious about this. We are just attaching, say, nine more numbers to each point in space besides the three which tell us where it is. The others tell us about the forces on a body there. These numbers form a 12-dimensional manifold. But that does not mean that space has 12 dimensions.

Similarly we can think about a four dimensional manifold of space and time. We are using four numbers when we say that an aeroplane was 12 miles north, 4 miles east, and 3 miles up at 4.23 p.m.

This way of looking at events is very useful, because two observers in relative motion, for example one in a moving train and one on a platform, do not quite agree as to what events take place at the same time, any more than they do as to what objects are at rest, but one can fit all their private views of the universe into one public space-time, which is the same for everyone.

So much for the uses of many dimensional geometry. Now for its abuses. There are people who explain ghosts and other marvels by a fourth dimension of space. They say things come into our space from it, and so on. If you answer that no one can see it or feel it, they reply that one can't see or feel radio waves, but one can pick them up with a receiver.

Now theories about invisible and intangible objects are only worth anything if they enable one to predict.

I believe in radio transmission because this theory lets me predict that by making the correct adjustments to a certain instrument I shall hear a certain programme. The predictions are not always right, but often enough.

Some people certainly see ghosts, that is to say, they see people who aren't there. I have been a ghost myself. I appeared to my grandmother while I was sitting on a lawn 20 miles away. If I had been dying or dead the vision would have got more publicity.

I cannot see how the existence of a fourth dimension of space would help us in the least to understand how a man's wife can see him when he is dying 1,000 miles away, or with the help of a medium after he is dead, supposing either of these things to be true. If there is a fourth dimension, we know enough about its geometry to say that any path through it between points in our space is longer than the shortest distance in three dimensions.

And above all no one has explained why objects from the fourth dimension appear in our space so very rarely. So I shall go on lecturing about the geometry of many dimensions when I want to, but I shall no more believe in the real existence of a fourth dimension of space than I believe that an economic depression is a real hole, or that the peak of gas consumption on Sunday mornings is a real peak which I could climb on a holiday.

35

Relativity

I have been asked to give a brief account of Einstein's theory of relativity. I cannot do so adequately in 1,200 words, nor without mathematical symbols. Nevertheless I shall try my best.

Newton's physics already go halfway to Einstein, and if you really understand Newtonian relativity, you will not find it so hard to grasp the more modern theory. Imagine a compartment in a train going very smoothly at 40 miles an hour. A man sitting in it drops a stone. To him it seems to fall vertically. That is to say it only moves vertically and not horizontally in relation to the other things in the compartment.

But if a man on a station platform watches the falling stone, he says it is moving horizontally at 40 miles an hour, besides its vertical movement. To him the stone appears to fall in a parabola, and to have moved forward for 29 feet while it fell the first 4 feet. Both these men are right in the sense that their thinking is this-sided, and will enable them to calculate correctly at what distance the stone will be at any time from objects on the train or the platform respectively.

The mechanics of a system are independent of the speed at which it is travelling, provided this speed remains unchanged in magnitude or direction.

Your feelings, or a plumbline hung in the compartment, are affected if the train slows down, accelerates, or goes round a curve. But no mechanical observations inside the train will tell you which way the train is going. Nor will they tell you anything about the speed with which the station platform is moving round the earth's axis, or the much greater speed with which the earth is moving round the sun. You can measure the changes in these speeds by a gyrostatic compass or in other ways. But you cannot measure the speeds themselves.

Clearly unless there is some way of detecting absolute rest and measuring absolute speed, space and time are mixed up in a curious way. For one man says the stone is moving in a straight line, and the other in a curve, and one seems to be as right as the other.

It was long thought that there was such a way, namely by means of light. One of two things might have been true. Light might have moved at a constant speed relative to absolute space. If this were so the speed of light moving eastward relative to a measuring apparatus would be different at noon and midnight on account of the earth's spin on its axis. And the difference between the speeds in January and July would have been still greater.

Or else the speed of light from an object moving towards us, for example Venus as an evening star, would be greater, relative to objects on the earth, than when it is moving away. Neither of these is true. All kinds of light move at the same speed relative to objects on the earth, regardless of the time which the measurement is made, or the object which sends out the light.

Since radio waves behave like light, no method has been discovered to find out how fast an object is moving, and according to Einstein's theory there is no way of finding it out. In fact, the question is a meaningless one. We can only find the speed of one thing relative to another thing. This is quite a simple idea, but it leads to very odd consequences. For one thing, measurements of moving objects are slightly affected. The moving train is slightly shorter when measured by a man on the platform than when measured by a man on the train with the same foot-rule.

Measurements of time are also affected. A watch in the train will record slightly less time in a given interval than a similar watch on the platform. The differences are much too small to measure at present when the relative motion is as small as that of a train relative to a platform. But unless they are there, there is a way of determining absolute rest and motion.

Again, different observers would disagree as to what events happened at the same time. The disagreement would only be measurable if the observers were moving at enormous speeds relative to one another, speeds which were an appreciable fraction of that of light, but there is no way of getting round it.

One cannot in practice get two observers moving fast enough relative to one another to make such measurements. But one can get small particles moving quickly enough to show that their mass and weight increase with their speed, as they should on the theory of relativity.

In fact, the calculations as to the energy liberated by atomic fission are based on the theory of relativity. For some of the weight of uranium or plytonium is due to the high speed of the particles inside their nuclei, which get out when there is an atomic explosion.

So far, everything is comparatively simple. I have described what is called the Special theory. But this theory does not deal with the mechanics of a system whose speed is changing. This is the province of the General Theory of Relativity. Almost all physicists agree with some of this theory, but not many agree with all of it, and indeed, it is far from complete.

If we accept the theory of relativity we have to admit that space and time are nothing absolute, but merely ways in which we describe the relations between different events.

This has often been taken to be idealism, but it is not. The falling stone in the railway carriage moves in a straight line relative to the carriage and a curve relative to the platform. This does not mean that the stone, carriage, and platform are not all real.

It does mean that our ordinary notions of space and time are not sufficient to describe the relations of three objects in motion relative to one another. Why should they be so?

Geometry is derived from two Greeks words, *geo* meaning land, and *metron* meaning a measure. It was originally a set of rules for measuring land, or surveying, and for this purpose it works extremely well. It now turns out that it is not sufficient to describe what happens if the surveyor is moving over the land, and a train is moving over it also.

One cannot deal with such a system without using a system of measurement which includes time as well as space, and is naturally more complicated than geometry.

That does not mean that the surveyor, the land, and the train are not real.

It does mean that philosophers like Zeno, Gegel, Marx, and Engels were quite right in saying that our ideas about motion were self-contradictory. However, they had not the special knowledge needed to resolve the contradictions.

Einstein has only done so partially. Very small distances and times are even more troublesome than very high speeds, and the combination of them is still worse. However, the difficulties are gradually being overcome. And each step in overcoming them leads to a greater control over matter.

36

Matter and energy

When an atom of uranium or plutonium explodes it breaks up into fragments which can be weighed, and also gives out a flash of radiation like an X-ray, but with a smaller wavelength. The combined weight of the fragments is distinctly less than the weight of the original atom, and the weight lost is proportional to the energy which appears as the motion of the fragments and the flash or radiation.

When the particles have been brought to rest and the flash has moved away there is a loss of weight. There is also a loss of mass. By the mass of a body we mean its tendency to remain in uniform motion or at rest even when there is no friction to stop it moving.

It is much harder either to move or to stop a lump of lead suspended from a rope than a football of the same size but a hundredth of its weight.

Mass is harder to measure than weight, but it is more constant. The lead and the football weigh less at the top of a mountain than at sea level, as you can see with a spring balance. This is because they are farther from the earth, so it pulls them less. But their masses are the same.

In any particular place mass is proportional to weight. Physicists prefer to think in terms of mass just because it does not depend on the earth's pull. However, it is not an absolute property of a body. A rapidly moving body has more mass, and more weight, than the same body at rest. But at ordinary speeds this difference is quite negligible. A body has to move at 71 per cent of the speed of light before it doubles its mass and weight.

Even a V2 weighing ten tons when at rest only gained about a tenth of a milligram when moving at 2,000 miles per hour. In fact, energy has mass and weight, whether it is the kinetic energy of a moving body or the potential energy of a coiled spring, a compressed gas, or an explosive compound. Your watch is heavier when you have wound it. It is also more massive. For example, it would be harder to start a sprint when wearing it. But the difference is immeasurably small.

For the life of me I cannot see anything particularly surprising, let alone paradoxical or mystical, in such facts as these.

A spinning body, for example a bicycle wheel, puts up a resistance if you try to twist its axis. A moving body or a body with a lot of potential energy puts up a resistance of another kind. It is harder to alter its state of motion or rest, and also harder to prevent it from assuming its natural condition of falling in a gravitational field such as the earth's. In other words, it has more mass and more weight.

The differences in mass between different kinds of atoms enabled physicists to calculate fairly exactly the energy set free when one kind is transformed into another, and thus the amount of energy liberated by an atomic bomb. It is only when the amount of energy liberated by a piece of matter is very large that there is any measurable loss of weight.

The percentage loss when we strike a match is too small to measure, provided we weigh the gas produced as well as the ash. On the other hand, a fair fraction of 1 per cent of the weight of an atomic bomb is lost.

If we like to say that mass is the amount of matter in a body, then we can say that matter is transformed into energy. But as the number of elementary particles, such as protons and neutrons, appears to be unaltered in an atomic transformation, it seems simpler merely to say that the system has lost some mass.

The mass is not annihilated. Even light has mass, as Lebedev showed when he blew pollen grains in a vacuum out of their path with a beam of light. It also has weight, as Eddington found when he measured the deflection of starlight passing near the sun during an eclipse.

Some readers will complain that I am merely talking about a form of words. My answer is that we use words and other symbols to help us to understand the world, and we had better use the clearest language of symbolism that we can.

We can be sure that reality will turn out to be more complicated than any form of language of algebra that we can invent. But we don't help ourselves by complicating things still further.

Scientists who talk in paradoxical phrases like the transformation of matter into energy have a certain excuse. One can sometimes shock oneself or one's hearers into accepting a new idea by using them.

It is the same in politics. The phrase 'property is theft' had a certain value in shocking people into realizing how grossly unjust was the claim of individuals to own hundreds of square miles of land, or millions of pounds of shares.

But as a generalization it is simply untrue, and leads to false historical theories such as Herr Duhring's 'force theory' to which Engels devoted so much attention.

In the long run such phrases do more harm than good. At the present time it is of the utmost importance that ordinary people should understand the development of science. We shall not help them to do so if we use language which blinds them to the fact that Nature is natural.

37

How we measured the atom

George Sinfield, the Sporting Editor of the *Daily Worker*, recently asked me how we knew the size and weight of atoms. I told him the answer would take about twelve hours. But I must try to give some of it in this article.

Over a century ago Dalton laid the foundations of our knowledge. Hydrogen and oxygen form two quite different compounds, water and hydrogen peroxide. In water there are eight parts of oxygen by weight for each one of hydrogen; in peroxide there are sixteen. It is a reasonable guess that oxygen is made up of atoms, either eight or sixteen times as heavy as those of hydrogen. When a few thousand compounds had been analysed in this way, the relative values of the atomic weights became pretty certain.

But nobody knew how small the atoms were. And some philosophers said they were only conventions to help our thinking, like parts of speech in grammar or the decimal system in arithmetic. Chemical changes occurred as if there were atoms, but we could never know what matter was really made of. You can read quotations from them in Lenin's *Selected Works*.

The first measurements which led to anything like the right figure were made by J. J. Thomson at Cambridgc about 1896 on the electron, the smallest electrical charge. Later on, Millikan measured it more accurately by a simple method. He made a spray of very fine oil drops. Some of them had an electrical charge which could be calculated from the speed at which they moved between two charged metal plates. The charge on a drop was just one, two, three, or some whole number of Thomson's units, never a fraction such as one-fifth or two and a half.

But, according to chemical theory, one of these electrical units went through a wire for each silver atom deposited in electroplating. So the number of silver atoms in an ounce could be calculated, and hence the sizes of all other atoms.

Radium told the same story. It shoots out nuclei of helium atoms with such speed that a single one makes a visible flash when it hits a screen of the right sort of material. The flashes per minute produced by a small amount of radium can be counted, and also the amount of helium gas produced by a much larger quantity in a month. Thus the size of a helium atom is known.

It turned out that the distance between neighbouring atoms in a solid is always about a hundred millionth of a centimetre, generally rather more, but never as much as five times more. In particular, the distance between layers of atoms in a crystal can be calculated. Now, if a series of very fine lines, say, a ten-thousandth of an inch apart, are ruled on glass it has long been known that this glass reflects light in a different way. White light gives a rainbow effect, as with a pearl or a pigeon's breast. But light of a single colour—for example, the yellow light from a sodium street lamp—is only reflected in certain directions.

In these directions the difference between the lengths of the paths travelled by light reflected from two neighbouring lines is just one, two, three, or some other whole number, times the wavelength of the light, and successive waves help one another. In this way, the wavelength of light can be measured.

Now, a crystal behaves in this way to X-rays, reflecting those which strike it at a particular angle. When the Braggs discovered this, they were able not only to calculate the structure of crystals, but the wavelength of X-rays, given the size of an atom. The calculated wavelength agreed well with other properties of X-rays—for example, the voltage in the discharge needed to produce them.

But the finishing touch was put by Siegbahn, who used soft X-rays—that is to say, rays of long wavelength and low penetrating power. He could reflect the same rays at a steep angle from a very finely divided ruled grating, and at a glancing angle from a crystal. The figures which he calculated for the distances between atoms agreed to 1 part in a 1,000 with the earlier figures.

When soap is added to water, it affects its surface properties much more than its bulk properties. The soap molecules form a layer on the surface of the water. Most water has a surface layer of some impurity. Indeed, the first person to make really clean water was a Miss Pockels about 1895. Clearly no layer of a substance can be less than 1 molecule thick. So by cleaning water thoroughly, and then seeing how far a drop of oil will spread, we can measure the thickness of 1 molecule.

If the surface layer consists of protein—for example, egg-white—it can be skimmed off on to a clean glass or metal plate and dried. Then this is done again, and so on. In this way Miss Blodgett got a layer of dried egg-white 1,764 molecules thick on a metal plate. You notice how women beat men on this very fine work. She stripped the layer off, folded it up, measured it with a micrometer such as is used in fine gauging, and found the same thickness, within 1 per cent, as that given by X-ray measurements.

The size of molecules, and therefore of atoms, can also be calculated from the rate at which they settle in a very powerful centrifuge, such as I described in an earlier article, and in literally hundreds of other ways, which all agree pretty well, and give a coherent account of the structure of matter.

It is such a convincing account that if anyone produced evidence which overthrew it I should certainly give up writing for the *Daily Worker* or supporting its policy. For if all this practical and theoretical work was meaningless, scientific thinking would be no use, and among other scientific thinking, that of Marx, Engels, and Lenin.

And if that is so, I may as well start believing the BBO bulletins, sell my articles to Lord Rothermere, and spend the money in night clubs. But as long as I believe that one can get somewhere by rational thinking and action, I shall do my best to persuade others that this is so.

PART II

Essays from the Indian period (1957–64)

In 1915, Haldane was sent to the British Military hospital in Simla (India) to recuperate from wounds received in Iraq in the First World War. In his 'Autobiography in Brief', Haldane wrote: 'I determined to come back as soon as I could associate with Indians on a footing of equality.'

In the 1950s, Haldane visited India as a guest of the Indian Science Congress Association and moved to India permanently in 1957. He stated several reasons for this. Both Haldane and his wife, geneticist Helen Spurway, were offered research positions at the Indian Statistical Institute in Calcutta. Politically, he had been feeling alienated from the official policies of the British Government for several years. The final outrage (as he saw it) was the 'Suez crisis', when an Anglo-French attack was launched against Egypt in 1956. He had been a Marxist and an admirer of the Soviet Union for many years but that too ended in a bitter disappointment when Lysenko, supported by Stalin and Khruschev, denounced and suppressed Mendelian genetics. That seemed to have left only one choice, as far as Haldane was concerned. As Kingsley Martin noted in the *New Statesman*, 'The balm of Sanskrit philosophy assuaged the burns of Marxism.' On the positive side, he admired Nehru's neutral stand in a world polarized between the United States and the Soviet Union. Perhaps even more importantly, he was eager to build a research school to carry on several projects on Indian plant and animal species. The offer from Calcutta thus came at the right time, especially since he was soon to be superannuated from University College London.

Haldane gathered around him an active group of younger scientists (including myself) in Calcutta. Many publications resulted from our research in several disciplines, including human genetics, animal behaviour, biometry of plant and animal species, agriculture, population genetics, and mathematical genetics. He became an Indian citizen in 1961.

When political differences with Mahalanobis arose, in 1962 Haldane moved to Bhubaneswar to found a new laboratory of Genetics and Biometry. He died of cancer on 1 December 1964. One of his last acts was to write an entertaining poem on cancer.

Essays of the Indian period discuss a whole range of subjects, including non-violence, statistics, mining, space research, astronomy, geology, evolution, and physiology. An interesting view of Darwinian evolution from the Indian perspective was presented by Haldane. In America and Europe, it is realized that Darwin's greatest achievement has been to show that biological evolution has occurred and humans are a result of that evolutionary process. This realization was an important event in the intellectual life of Western countries because Christian theologians made a sharp distinction between humans and other living beings. But in India and China such distinction has not been made, and several Eastern religions in fact recognize that animals have rights and duties. Consequently, Darwinism is readily accepted in these countries.

Haldane's essays were read widely in India. He rapidly became well known to the lay public. Photographs and cartoons of Haldane in his Indian attire were frequently seen in the Indian newspapers, and he used to receive a lot of mail from readers, which he answered promptly. He enjoyed the pantheon of Indian religions and philosophies. It is quite correct to say that Haldane enjoyed India and India enjoyed Haldane.

38

Some statistical adventures

Statisticians are not usually very adventurous people, perhaps because they are more interested in means than extremes. However, every six months the N S S* must provide adventure for a few workers who find that they have to go to remote villages in the mountains of Himachal Pradesh or the jungles of Mysore. My own work has been only statistical in part. Nevertheless I began to contribute to statistical research at a very early age. My father invented a number of methods for physiological measurements. Among these was a haemoglobinometer with which one could estimate the amount of the red pigment haemoglobin in human blood. A drop is quite sufficient, but 60 years ago British people were as averse from giving blood as some Indians are now. As 'Student' had not yet published his great paper, my father did not know how many subjects were needed before he established with reasonable certainty that adult bloods contain more haemoglobin than those of children. I think I was the first child whose blood he sampled. But there was some difficulty in obtaining enough children, and my sister was heard to remark to another little girl in the street. 'You come in here, my father wants your blood.'

Later on I contributed numerical data on other physiological characters. But many of my father's experiments were too drastic to be done on large groups. So were some of my own. For example I ate about 15 grams of salt before going to bed, with no water. By the morning my kidneys were concentrating chloride to about .31 molar.

* National Sample Survey of the Indian Government.

On taking another 15 grams in the morning the concentration only rose to about .33 molar. Only four or five people were tested in this way, but their concentrating power varied from under 0·3 molar to over 0·4. I do not know whether I shall be able to obtain volunteers for this form of *tapas* in India. Intense thirst is not pleasant, and it requires a certain concentration of mind to resist it.

During the 1939–1945 war I was engaged in physiological work for the British Navy. Unfortunately naval officers do not understand statistical principles. They asked me some questions which I could answer, for example, 'How much compressed oxygen and soda-lime mixture are needed by two men if they are to keep alive in a miniature submarine for 48 hours, doing very little work?' This question can be answered because the oxygen consumptions of different men at rest are not much more variable than their weights. By staying shut up with a colleague for the requisite time I convinced them that it would be safe for others. They next asked me how long a man can safely breathe oxygen at various depths under water. This knowledge was needed for the men who proposed to put bombs under the German battleship Tirpitz, wearing self-contained diving dresses. Oxygen is poisonous at high pressures, causing convulsions which can be quite unpleseant and are followed by unconciousness for five minutes up to half an hour. Sometimes there are warning symptoms, sometimes not. We soon found that the time before the onset of convulsions fell off rapidly as pressure increased. At depths over 40 metres they often occur in less than five minutes, and may be so violent, at least in my own case, as to break bones.

But different people seemed to vary a great deal in resistance to oxygen poisoning. I say they seemed to, for the navy not unnaturally did not wish to frighten men who had volunteered for a dangerous job by giving them convulsions. To determine whether one could trust the result of a single test we did a series of 17 experiments on my wife who breathed oxygen at a standard pressure until she had a convulsion, or she or the man observing her thought she was going to have one. The times which she lasted varied from 13 minutes, ending in a convulsion, to 92 minutes, with nothing worse than twitching. Only one sailor beat this. My wife's times had very positively skew distributions.* Altogether she has had five convulsions from high pressure oxygen, with no worse injury than dislocations of the jaw. It was clear that one could not give a definite answer. We suggested that after a previous test to weed out the most sensitive men one could cut down the risk of a convulsion, which would mean almost certain death under water on active service, to about 1 per cent. However, the British naval authorities were not at all satisfied with this answer.

You see that some kinds of statistical investigation involve some discomfort and even risk. For I must admit that I was never quite sure that my wife or I would be pulled out in time if we had a convulsion under water. This Institute is surrounded by objects which are a challenge to statisticians, and whose investigation will involve some risk. I refer to the palm trees of various species. It may be possible to measure the yield of a cocount palm (*narikel*) without climbing it. But some day, and the sooner the better, an attempt must be made to breed for high yield. Now if we plant the nuts from a tree we know their mother. But we do not know the father, that is to say the tree which provided the pollen to fertilize the nut. Statistical work on the inheritance of yield can only begin when a geneticist can climb the coconut trees to make the required matings. I at least would feel safer 30 metres under water than 30 metres above ground on a swaying palm.

In my opening sentence I wrote that statisticians are usually more interested in means than extremes. I am exceptional. I have just taken up the problem of the distribution of maximal values in samples

* For the benefit of non-statisticians, I explain that this means that a few times lasted much more than the average, and many a little less.

where Fisher left it thirty years ago. And I think I have shown that the maximum value of the square of a normal random variable has a much simpler distribution than the unsquared maximum. Such intellectual adventures are doubtless a poor substitute for having convulsions under water. But they are better than nothing.

39

Some reflections on non-violence

I am a man of violence by temperament and training. My family, in the male line, can, I think, fairly be described as Kshattriyas. Before 1250 our history is fragmentary. From 1250 to 1750 we occupied a small fort commanding a pass leading from the hills to the plains of Scotland. Our main job was to stop the tribal peoples of the hills from raiding the cattle of the plainsmen; but perhaps once in a generation we went south to resist an English invasion, and at least two of my direct ancestors were killed while doing so. Even when Scotland had been united to England by a royal marriage and the tribals had been pacified, the tradition persisted. When I was a child my father read to me Scott's 'Tales of a Grandfather,' which are legends of the warlike exploits of the Scottish nobility, and trained me in the practice of courage. He did not do so by taking me into battles, as his ancestors might have done, but by taking me into mines. I think he first took me underground when I was 4 years old. By the time I was about 20, I was accompanying him in the exploration of a mine which had recently exploded, and where there was danger from poisonous gases, falls of roof, and explosions. So when in 1915, I was first under enemy shell fire, one of my first thoughts was 'how my father would enjoy this'.

I find many of the virtues and vices of the heroes of Indian epics quite intelligible and even sympathetic. The second word of the Gita, 'dharmakshetre', gives an exact description of my feelings when I went to the trenches for the first time in 1915. I was well aware that I might die in these flat, featureless fields, and that a huge waste of human values was going on there. Nevertheless I found the experience intensely enjoyable, which most of my comrades did not. I was supported, as it were, on a great wave of dharma. The European Kshattriya, or knightly virtues include a detestation of various kinds of meanness, and a hatred of violence against the defenceless. The European knightly vices include an addiction to gambling. I understand Yudhisthira's point of view. A Kshattriya should never feel secure. His dharma implies that he must be prepared to risk his life, and lose it if necessary, at a moment's notice. He must therefore be prepared to risk his property. I confess that I have less sympathy with his staking his wife and his brothers.

In the war of 1914–1918 I was on several occasions pitted against individual enemies fighting with similar weapons, trench mortars or rifles with telescopic sights, each with a small team helping him. This was war as the great poets have sung it. I am lucky to have experienced it.

We have now to consider two facts. The Gita, which is an exhortation of Arjuna to violent conduct, was the favourite poem of Gandhi, the great exponent of non-violence. War has changed its character completely in my lifetime. Modern war has two principal forms. One form is characterized by the wholesale massacre of defenceless civilians with atomic bombs and other weapons. The other, which is going on in Algeria, Malaya, Kenya, Cyprus, and other regions, is characterized by the use of ambushes and individual murder by the less well-armed side, and the killing of prisoners and the enslavement of whole populations by the more strongly armed side.

Modern war does not evoke any of the Kshattriya virtues except courage. But yet these virtues are absolutely needed in modern life, as Gandhi saw. The contradiction is of course latent in Hindu mythology. Not only Rama and Krishna, but even Buddha, the great preacher of non-violence, were Kshattriyas. But Parasurama, the son of Jamadagni, another avatar of Vishnu, had devoted his life to the extermination of that caste. How then can we combine the Kshattriya virtues with non-violence? Gandhi gave one answer to this question. There are other answers, quite compatible with Gandhi's answer, but in different spheres. Gandhi was always concerned in struggles between human groups. He did his best to eliminate violence and hatred from them.

There is another kind of struggle. I quote St Paul's letter to the Ephesians, the translation from the Greek (from memory) being my own : 'For our struggle is not against blood and flesh, but against first principles, against powers, against the *lokapalas* of the *kaliyuga*, against the spiritual sources of evil in the heavens.' I translate his word *kosmokrator* or world governor, as *lokapala*. The phrase translated as *kaliyuga* means literally 'the darkness of this age.' I think that the notion of the *lokapalas* had reached western Asia from Buddhist sources in St Paul's day.

Some of us struggle against the natural forces which in India are too often worshipped as minor deities, for example cholera and smallpox. My father was mainly concerned with such matters as the ventilation of factories and mines, which is important both in safeguarding health and preventing explosions. When he wished to investigate why men died after colliery explosions when they had received no physical injury, he first examined dead men and horses after underground explosions, convinced himself that they had died of carbon monoxide poisoning, and then proceeded to poison himself with this gas. That is to say he breathed a known amount of it until he had fallen over unconscious, and a colleague pulled him out of the gas chamber. In this way he found out how long it takes for a given amount of this gas to overcome a man. He also found that small birds are overcome much more quickly than men (and recover much more quickly). He was, however, averse to experiments on animals which were likely to cause them pain or fear (carbon monoxide poisoning causes neither). He preferred to work on himself or other human beings who were sufficiently interested in the work to ignore the pain or fear. His experiments on the effects of heat could perhaps be called *tapas.* He found that he could live in dry air at 300°F. At about this temperature his hair began to singe when he moved it.

But I do not think his motivation was that of an ascetic practising *tapas.* He achieved a state in which he was pretty indifferent to pain. However, his object was not to achieve this state but to achieve knowledge which could save other men's lives. His attitude was much more like that of a good soldier who will risk his life and endure wounds in order to gain victory, than that of an ascetic who deliberately undergoes pain. The soldier does not get himself wounded deliberately, and my father did not seek pain in his work, though he greeted a pain which would have made some people writhe or groan, with laughter. I think he would have agreed with the formulation that the *atman* or *buddhi* in him was laughing at the *ahamkara.*

I have tried to imitate him. I have drunk or breathed considerable amounts of various poisons, certainly more than half the fatal dose in some cases, and have done similar experiments on other human volunteers, including my wife. For this reason I feel a certain annoyance when I am excluded from a temple of Siva, who, according to a well-known legend, drank poison to save the other gods. If Siva exists, he may be more pleased by such an action than by the recitation of a lakh* of *mantras.*

I believe that this non-violent approach to experimental biology is a fruitful one. I do not condemn those who do experiments on animals which involve their death, or even moderate suffering. But I have never done an experiment on an animal of a kind which I have not previously or subsequently done on myself ; and I hope I never shall. I have dissected dead animals. But I have left instructions that my own body should be dissected by medical students. And if I die in

* 1 lakh = 100,000.

India I hope some future Indian doctors will have the unusual experience of dissecting a European.

One great advantage of working on oneself or a friend is that far greater accuracy is possible than is usual in experiments with animals. If an animal is in pain or fear, for example, its heart is likely to beat faster and its rate of breathing may also increase. This will make it impossible to measure the effect of a drug on its pulse rate or breathing with great accuracy. But when one has done a number of experiments on oneself, one can do them on animals with some confidence that the results will be as accurate as if they were done on men.

When we were in India in 1954 my wife and I did a number of experiments of this kind on three *koi* fish. When these fish are put in foul water containing little oxygen they swim up to the surface and breathe air. We teased our fish in various ways by altering the composition both of the gases dissolved in the water in which they swam and the air above it which they breathed. I don't say these fish never suffered at all. To judge from my own experience they may have had severe headaches for some minutes. But they were certainly not seriously injured, for all three of them are alive and well in London today.

I am not at all a saintly person. I have killed animals, and eat meat, though not very much. Since writing this, I have ceased to eat meat or fish. My attitude to animals is more like that of Yudhisthira. He had killed and eaten plenty of deer. But when he was asked to enter *Svarga* leaving the dog which had been his companion on his last pilgrimage to die on the mountain, he found, perhaps to his surprise, that this was something which, as a *Kshattriya*, he could not do. No more could I or my wife have given the *koi* fish which we had watched for two months to someone else to kill and eat.

I want to urge that this kind of attitude to animals should commend itself to Indian biologists. Unfortunately it is rather rare. I think that it should be the rule. But of all the Indian biologists whom I know the man who comes nearest to it is a Muslim, your great ornithologist, Mr Salim Ali. He is prepared to admit that he has occasionally shot birds, but he greatly prefers studying them when alive.

Frankly I regard his attitude as a challenge to Hindu biologists. There is a very great opening for non-violent biological studies in India, and, what is important, they require no complicated apparatus. Let me give some examples of what could be done. Do your song birds sing their songs if they are brought up from the egg by human beings? Or do they have to learn their song as a human being learns a language? In Europe and North America we know that some birds must learn their song, while others produce it untaught, as some mythical Hindu characters were born with a knowledge of the Vedas. If they have to learn, some species learn from their fathers. Our English robin sings very little while helping his wife to keep the eggs warm and to feed the children which emerge from them. But when the children are just learning to fly he bursts into song again for a few weeks. This may be an expression of paternal pride, but it gives his sons the chance to learn from him. Other birds do not learn till they are nearly a year old, when they learn from other males. To bring up young birds till a year old requires an aviary where they can fly about, and a great deal of devotion. The latter is common in India, but it is commoner among illiterate people than among biologists.

If ever I settle down in India I hope to continue the study of the almost non-violent branch of biology called animal genetics. If, for example, I want to breed ducks, the conditions in many parts of India, including much of West Bengal, are ideal. One would need some wire netting to prevent unwanted matings, and perhaps a balance, a tape measure, and a chart of colours, but no other apparatus. One would of course need land with small tanks, a certain amount of food, but above all, assistants who would take the greatest care of the ducks. Such work would certainly yield many facts of biological interest, and probably increase the egg production of your ducks. Genetics is not, of course, completely nonviolent. We breed a lot of flies in my laboratory, of a species which normally lives on sap from injured trees. If we let the surplus ones out in London they would starve to death. We therefore anaesthetize them and drown them in oil before they recover from the anaesthetic. I sympathize with, but do not share, the Hindu practice of non-violence to insects. They will die in any case, but they need not suffer. I am ashamed if I cause an insect suffering, but not if I kill it painlessly. It would be quite possible to practice insect genetics without offending the scruples even of a Jain. But it would mean liberating animals most of which would starve to death.

Gandhi was quite clear that men have a duty of non-violence to animals. And there is no reason why biological research in India should not be conducted on Gandhian principles. On the contrary, there is very good reason why much of it should be so conducted. If Indian physiologists were ashamed to do an experiment on an animal which they could do on themselves, Indian physiology would, in my opinion, be considerably more fruitful. The most effective method of mosquito control is not to kill mosquitoes, but to give them no opportunity of laying eggs. If you have too many cows, you should try to breed cows which will go on giving milk for more than a year after a calf is born, as some of our European breeds can. You will then have fewer cows, and more grass for each cow to eat. If your traction is largely mechanized in future you will want fewer male calves. My wife has a male fish who (as predicted before he was tested) has begotten only daughters. There does not seem to be any intrinsic impossibility in producing bulls who would do the same, though many lifetimes of human research may be needed before this is achieved.

Thousands of Indians die of snake bite every year. But the cobra is one of the most beautiful of animals, and I should be very sorry to have to kill one. It is just as possible to immunize the rural population of India against the poisonous snakes which live in their neighbourhood as to vaccinate them against smallpox. But the methods for doing so safely and on a large scale have not been worked out. This is a task for non-violent Indian biologists. Two or three of them might die of snake bite before the process was fully worked out. If so they would not have died in vain. At a later date it might be possible either to breed cobras whose poison did not kill men, or men who were not harmed by the bite of a cobra. Meanwhile I should like to meet even one Indian biologist who had immunized himself (or herself) against all snake poisons, and was prepared to answer a telephone call and remove an unwanted cobra or krait from a house without killing it.

Such ideas no doubt seem silly to many Europeans. If they seem silly to Hindus, this means that there is something badly wrong with modern Hinduism. In my opinion several things are badly wrong with it. In particular the love for all animals which is expressed in your scriptures and your art has been replaced by a series of formal prohibitions. Gandhi realized that if non-violence to human beings is to be effective, it requires both courage and intelligence. He had plenty of both. I have tried to show that courage and intelligence are needed if non-violence to animals is to be something positive, based in all cases on the love of men for animals, and in many, on the love of animals for men.

Politicians may not like this article. They may say that it calls for a diversion of effort from important to unimportant fields. I do not agree. In my experience kindness to human beings and to animals usually go together. Those who ignore suffering in animals find it easier to ignore human suffering, and conversely. I think that Indians who love animals are often perplexed because they do not see how to give practical effort to their love. I have made some suggestions, and venture to hope that some of them will be accepted. India has made many contributions to world culture. Perhaps the greatest is the ideal of non-violence. Europe's greatest contribution is the scientific method. If these can be married, their offspring may raise mankind to a new level.

40

Science and floods

Once more the monsoon of 1959 has brought disastrous floods, the worst being in Kashmir at the beginning, and now, near the end, in West Bengal. It is reasonable to ask what scientists can say about their cause and prevention. If one may give an epigrammatic answer, floods are a sign of immaturity both in landscapes and in governments.

The landscape of Kashmir is extremely immature. You cannot go far in the valley without seeing some of the peculiar hills called moraines. These are made of clay with large and small stones, and have been formed by glaciers. A glacier is a slowly moving river of ice which can carry down not only mud and gravel, but boulders as large as houses. These are deposited wherever the ice melts. The commonest kinds are lateral moraines at the edges of glaciers, and terminal moraines formed at their ends when the climate remains steady for some centuries, and the glaciers neither advance nor retreat, but deposit their loads in the same place for a long time. The beautiful meadow of Gulmarg, for example, is almost surrounded by terminal moraines. Now a glacier has nothing corresponding to floods. So when the glaciers of Kashmir began to melt twelve thousand years or so ago they left a landscape in which the valleys were not adapted to deal with a sudden rush of water. We have a few similarly immature areas in England, notably the Vale of York, where moraines laid down by ice hold up water which would drain away more quickly if valleys had been made through them by nature or by men.

The landscape of West Bengal is immature for other reasons. In the north the Himalayas are still rising, as is shown by the frequent earthquakes, and in the south the delta, to which the Ganga and other rivers such as the Damodar contribute, is still growing at the expense of the sea. The southern slopes of the Himalayas are so steep that the rivers can carry big loads of gravel which they deposit in the plains. These not only ruin fields, but cause the rivers to seek new courses.

The most homicidal of all rivers is the Hwang Ho in north China. Its upper reaches run through the fine dusty soil called loess, which was blown to its present position by winds during the last ice age. This is being eroded, and laid down as mud in the plains below. The river bed gradually rises above its surroundings, and naturally changes its course from time to time, drowning lakhs* of people. Indian rivers such as the Teesta do the same, but fortunately on a much smaller scale. The Po and its tributaries in northern Italy have the same tendency, but they have been controlled to a considerable extent, and Indian engineers would do well to study Italian methods of flood control. It appears that the dredging of several rivers in West Bengal, such as the Bhagirathi, has been seriously neglected.

If nature takes its course, these dangers will largely disappear in a few million years, or even a few lakhs of years. The Teesta and similar rivers will have brought down enough gravel to enable them to form relatively permanent valleys. The floor of the valley of Kashmir will have been raised, and the gorge through which the Jhelum** leaves it will have been deepened. The delta regions of Bengal will slope gently like the plains of Uttar Pradesh.

But human beings cannot wait. What can they do now? The first thing, no doubt, is to collect information and analyse it statistically. This does not appear to be done very adequately in India. In a leading daily newspaper of 5 October, I read the following sentences about the floods in the Damodar Valley, which raised the water level behind the Panchet dam. 'Previous statistics show that the probability of heavy rain so as to raise the level to 433 feet is once in 50 years and to raise the level to 435 feet is once in a thousand years. Both the figures were crossed this time.' If these figures were invented by an imaginative reporter he would be well advised to take up science

* 1 lakh = 100,000.
** Jhelum = Jhelum river in Pakistan and India.

fiction; if by a statistician he would be of more use to his country if he were carrying earth in a basket to make bunds.

If there had been continuous records of rainfall or stream flow in the Damodar Valley since the time of the battle of Kurukshetra, which is commonly thought to have been somewhat over three thousand years ago, we could make a rather uncertain prediction as to what was to be expected in the next thousand years, assuming that various conditions remain steady. In fact the longest series of rainfall records in the world is for Pavia in northern Italy, and goes back for only about 250 years. I should be surprised if any rainfall records in the Damodar basin went back much beyond 1900. If they do they do not help us. A heavy shower may deposit ten inches of rain in a small area during a day. This may not cause serious floods. But four inches in a day over a large area will do so. We have not got enough data for any part of India to make predictions about the worst flood to be expected in a century, let alone a thousand years.

With some kinds of statistics we can make predictions about a sample of a thousand on the basis of a sample of a hundred. This is so, for example, for heights of adult men. It is true for annual rainfalls in some places, but not in others. But even if it were true for rainfalls over several centuries it would not be true over longer periods. To take the history of only the last two thousand years in Europe we know that winters were a good deal colder in the time of the Roman Empire than today. Armies crossed rivers on ice which are now seldom even lightly frozen. From about AD 500 to 1000 the weather was warmer than now, and Greenland was colonized. Then came a colder spell of several centuries, and the last century and a half has been warmer. At the present time we are having unusual weather. The summer of 1959 is said to have been the warmest in England for two centuries.

This may be due to natural causes. And if so the general average of the weather may be changing. Or it may be due to the discharge of atomic bombs, whose full effects are not understood. Those who assert positively either that they have or have not affected the weather seem to me to be about equally unscientific. However that may be, the weather in the next century may turn out to be very different from that in the last.

The Panchet and Maithon Dams, like many others, are supposed to serve a triple purpose: hydroelectric power supply, irrigation, and flood control. At the beginning of the monsoon they probably do so. The water level in the lakes above them has fallen fairly low, and there is plenty of room to receive large amounts of water if there is heavy rain in the catchment areas which drain into them. The Kashmir floods were at the start of the monsoon, and a few empty reservoirs would at least have alleviated them. But during the monsoon these dams filled up, and their sluices had to be opened to avoid a possible burst. The same dam cannot be fully efficient for water shortage and flood control. The level at which the water is kept is a matter of policy which should be decided scientifically on the basis of such statistics as exist and of weather forecasts. I do not know how it is decided. I do know that the decision made this year was disastrous.

There are other methods besides damming rivers and dredging them by which floods may be avoided. One is by planting more trees in the areas of heavy rainfall, and protecting those which exist from destruction by men or goats. Planting is particularly needed in Kashmir. Another is by proper siting of new towns. According to the same newspaper which I quoted earlier, at Durgapur 'the whole area in the plant site is totally submerged'. Quite enough of our old cities are built on sites which are easily flooded. To build a steel plant round which a great city will grow up on such a site is a piece of stupidity, whether stupidity of Ministers or of civil servants it is probably impossible to discover.

Floods are a more serious danger in India than in western Europe, and Indian scientists and engineers will have to work out their own methods of preventing them. The Chinese are certainly tackling the matter in a very big way, but I do not know whether it is the most efficient possible way. And the Hwang Ho at least is decidedly unlike any Indian river. So I do not know if the Chinese can help us very much. I believe that Indian scientists can do more for their country and for the world by investigating the causes of Indian floods and the methods of preventing them than by much of the research which they are actually doing.

41

Colliery explosion

There has been a serious colliery explosion at Chinakuri in West Bengal, in which nearly 200 people were killed. Such explosions were common in the United Kingdom sixty years ago, but have been rendered extremely rare. The improvement has been due to three causes: scientific research, with some of which I was associated; regulations for safety; and education not only of engineers and mine inspectors, but of miners. It is idle to hope that safety will be achieved in Indian mines until the men working underground understand the sources of danger, and how to counteract them. And this improvement can only come as part of a general rise in educational level. Even that will not be sufficient unless the general education includes at least some elementary science.

An explosion is a burning which happens almost instantaneously through a large volume. In most cases the burning is the combination of oxygen atoms with other atoms. In a solid explosive such as gunpowder the oxygen is in potassium nitrate, from which it comes off very easily, and later combines with carbon and sulphur. In an explosive such as trinitrotoluene the oxygen and the carbon and hydrogen with which it unites are originally in the same molecule, but separated by nitrogen atoms. However, in a colliery explosion things are different. The oxygen is in air, and what is burned is either a gas called methane, or coal dust. So to prevent explosions one should take two different kinds of precaution. Firstly, one should prevent the accumulation of gas or dust in sufficient amounts to cause an explosion. And secondly, one should avoid anything which might start an explosion if gas or dust is present. In the same way if one has a gun or pistol in the house one should take two precautions. Firstly, the gun should not be left loaded; and secondly, it should never be pointed at a human being, even if one is sure that it is not loaded.

A gas or dust explosion is dangerous on account of the heat generated by the rapid burning. This can scorch a man's skin, but a greater danger arises from the rapid expansion of the air caused by the heat. This pushes everything in its path, as the explosion of the mixture of air and petrol vapour in the cylinder of a car engine pushes the piston. Both gas and dust explosions use up the oxygen of the air and generate a poisonous gas, carbon monoxide. Finally an explosion may start a fire. This seems to have happened at Chinakuri.

Methane, which is called firedamp in England, comes out of the coal in many mines, and in such mines 'naked' flames such as candles are forbidden. Either electric lamps are used, or an oil lamp is used with a glass chimney and wire gauze screens where the air enters below the flame and leaves above it. A flame can no more get through such a screen than a mosquito can get through the gauze screen of a railway carriage. When I was a boy my father gave me what Gandhi would have called basic education in science. I first went down a mine at the age of about four, though I did not go down a coal mine till later. I remember being taken into an abandoned part of a mine where there was no ventilation, and gas had accumulated. Methane is lighter than air, and collects near the roof. We crawled along the floor to avoid it. Then my father told me to raise my safety lamp. It filled with blue flame and went out. If it had been as large as a railway compartment the explosion would have gathered so much force that it would have shattered the glass and ignited the gas round it. Then my father told me to stand up. As there was very little oxygen in the firedamp near the roof I soon fell down again. However I had learned some chemistry.

Mixtures of gas and air in suitable proportions are rather easily set alight. A match can do so. So can a spark from an electrical apparatus, or perhaps even from a nail in a man's boot. In England a very common cause of explosions was what is called a gob fire. When the coal has been taken out of a seam the props which support the roof are withdrawn and the roof comes down. The crack which may remain between the roof and floor is called the gob or goaf. If too much coal dust and fragments are left there, and there is a draught through the gob, this waste coal may heat up and finally smoulder at a red heat. Such a gob fire can set off a gas explosion. Gob fires can sometimes be detected by smell long before any smoke is visible, but chemical methods are much more reliable.

Dust explosions are not so easily started. A cloud of coal dust must be quite dense before it explodes. Such dense clouds can be produced by wagons loaded with coal, by coal cutting machinery, and otherwise. More usually, however, coal dust is stirred up by a small gas explosion, and then set alight. Dust explosions can be more violent than gas explosions. I once saw an experimental one which tossed pieces of boiler as large as carts for 600 yards or so; and though the local newspaper was dissuaded from reporting this fact (for the English press is no more incorruptible than the Indian), an earthquake was reported six miles away.

Considerable care is taken in British collieries to prevent the accumulation of dust. Roads, as the tunnels between the shaft and the coal face are called, may be watered, limestone dust may be mixed with the coal dust, and so on. This is particularly necessary in mines where explosives are used. Although explosives used in collieries are tested for the absence of a flash, these tests are not quite infallible, and extra precautions are needed to prevent dust explosions. From the newspaper account of the Chinakuri explosion which I have seen, it looks as if the explosion of a small amount of gas might have started a much larger dust explosion. Enough gas to kill more than one or two men can only accumulate through very gross negligence. But a moderate amount may suddenly come out of the coal, and if it is accidentally ignited, will explode. However, if proper precautions are taken, it will not start a dust explosion.

The explosion only lasts for a second or less, but is extremely hot. So exposed skin is scorched. But though the clothing may be charred, the skin below it is seldom much damaged. Unfortunately in India miners wear fewer clothes than in Britain, and are likely to be burned over larger areas. Death by carbon monoxide poisoning is one of the least painful deaths that one can imagine. After a brief period of confusion one loses consciousness. Mine rescue teams are equipped with breathing apparatus which give full protection against it. It has been proposed that all coalminers should carry such apparatus. But this proposal has been rejected, as the apparatus is cumbrous. I am glad to read that a rescue team saved several lives at Chinakuri. Mine rescue is one of the noblest of human occupations, combining the honour and risk of the soldier with complete non-violence. The British teams have a magnificent record of courage, even though in the last few years they have had little chance of showing it.

No doubt there will be an enquiry into the Chinakuri disaster. Trade unions can play a most important part in such enquiries provided their representatives have a good knowledge of the necessary science and technology. Too often they are lawyers who know no more of chemistry and physics than I know of law, and are more concerned to make the witnesses called by the owners contradict themselves than to arrive at the truth. In such an enquiry it is no doubt important to find out, if it is possible, who was responsible for the explosion. It is much more important to prevent such explosions in future. This can only be done if the problem is regarded as a scientific rather than a legal question. But whatever precautions are taken I do not believe that Indian collieries will attain the high safety standards reached in Britain until the miners are sufficiently educated to understand not only what orders are given to ensure safety, but why they are given.

42

Deep mines

A number of miners have been buried in a very deep mine in Nova Scotia, a province of eastern Canada, and some have been rescued after more than a week. Press reports do not agree, but according to some of them the mine is a coal mine and just over 4,000 metres deep. This is a surprising but not impossible depth. In the days when I followed such matters the world's deepest mine was a Bolivian tin mine about 3,000 metres in depth. I have never been down more than a mile, which is about 1,600 metres.

However, even at this moderate depth some strange things were to be noticed. The shaft only went down vertically for 800 metres. The rest of my journey to the bottom was in a 'tub', that is to say a truck on a very narrow gauge tramline going down a steep incline, attached to a wire cable. The ventilation at the bottom of this incline was not too good, and it was extremely hot by English standards. I think even Indians would have objected to hard work under such conditions. The English miners who were working there wore nothing but a diminutive dhoti tied on with tape, and a pair of boots. When they stopped work at 'snapping time', for a small meal in the middle of the shift, they emptied the sweat out of their boots. The coal seam was less than a metre thick, so one had to wield a pickaxe lying on one's side, and although I was fairly robust, my efforts to do so, which only lasted for about ten minutes, greatly amused my companions.

There were three reasons why we were so hot. In the first place the temperatures of rocks increase steadily as one goes down. The gradients differ in different parts of the world, but one degree centigrade per 40 metres is a representative figure. The heat is being produced by radioactivity in the earth's crust. It leaks out extremely slowly. How slowly may be judged from the fact that the gradient in northern England is far from uniform. The temperature rises rather slowly in the first 300 metres or so, and then much quicker. This is only the case in countries which were covered with ice during the last ice age. This lasted for over 20,000 years, and cooled down the outer kilometre or more of the earth's surface appreciably. The last ice sheet disappeared from England about 12,000 years ago, but heat seeps so slowly that the rocks have not yet warmed up completely.

A second reason for the heat is that the air which comes down the downcast shaft is warmed up by compression. The air at the bottom of a shaft a kilometre deep has to support the weight of a kilometre more of air than it did when at the surface. It is therefore squeezed into a smaller volume, and heats up like the air in a bicycle pump. Finally in many mines the rocks heat up through the slow oxidation of minerals such as iron pyrites.

In very deep mines some form of artificial cooling is necessary, if men are to work as hard as they can. The simplest method is to compress air at the surface. It gets very hot during the process, but soon cools down. If it is led to the working face in pipes it cools down on expansion. This is one reason why compressed air is a favourite source of power in mines. Another reason is that it is far less likely to produce sparks than is an electric current, though a rapid air flow can produce a static charge. Sparks are of course a real danger in coal mines where they may start an explosion.

In a cave or a shallow mine the silence is frightening. After a few minutes' sitting still one begins to hear the sound of one's own breathing and heart beat, unless, as is often the case, water is dripping somewhere within earshot. But, in coal mines at least, at depths below 500 metres there is no silence. One hears a constant series of little cracks. At 1,500 metres these sounds are much louder, rather like rifle bullets passing in a battle. This sound is due to the rocks cracking under pressure. Coal mines are excavated in fairly soft rock, and as the coal, which usually lies horizontally, is removed, the roof gradually sags down, so that the miners work in a space which may be called a triangular prism. However, it is not only the roof which is under a strain. If one goes into a fairly deep coal mine after a strike which has lasted for a month or so, or which has been abandoned for a month or so after an explosion or a fire, one notices that not only has the roof fallen or at least become lower, but that the floors of the 'roads', as the tunnels leading to the face are called, have bulged up. The rocks are solid, but under the pressure of the strata above them they flow almost like a very sticky liquid. However, the flow is not steady, but in a series of little bursts. Apparently a big burst of this kind trapped the Canadian miners. The gold mines at Kolar in Mysore are deeper than any British coal mine, and no doubt tunnels close up if abandoned. But the rock is much harder than that of coal fields, and the gold bearing veins or lodes are nearly vertical. So the pressure effects must be less, and I do not know if the low levels are as noisy as a deep coal mine.

I have no doubt that our descendants will explore the depths of the earth as we are now exploring the upper air. But it will be as difficult for a man to go down 20 kilometres as to go up 200. I suspect that excavation will have to be done by apparatus under distant control, working in water or possibly some other liquid, under enormous pressure. This pressure will serve to balance the pressure of the rocks, and keep the shafts and roads from collapsing. The mere weight of a column of water will not be enough, for rocks are usually between two and three times as dense as water, and pressure will therefore have to be applied artificially to it. This is no more fantastic than the methods already employed in atomic power stations. The fact is that since the atomic industry is a new one, its workers are very much better protected than miners. For mining is a very old occupation. Moreover only a few mines in India are very deep, and the immediate task before the Indian Mineworkers Unions should be to see that the standards of safety current in Britain are enforced here, rather than to think of the fairly remote future.

But it is a curious thought that the mining methods of a century hence are probably being tried out today at the bottom of the ocean. Apparatus is being lowered which will allow an observer on the surface to see a picture of the ocean floor 8 kilometres below him on a television screen, to pick up objects so seen, to make boreholes in the ocean floor, and so on. All these have to work at pressures of about a thousand times that of the atmosphere. Their design calls for considerable imagination. But if we cease to regard miners as expendable material, they will probably be replaced by apparatus of this kind.

43

The sound of one hand

There are several schools of Buddhism in Japan. It may well be that some of these are as far from the sermon in the deer park at Kashi as the practice of the Church of England is from Jesus' sermon on the Galilean mountain. I shall not discuss this question. The Zen, or meditative school, is divided into two sects, of which the Rinzai claims that *samadhi* can be achieved by meditation on what is called a Koan, and probably mistranslated into English as a riddle. One of the seventeen hundred Koans is called 'the sound of one hand'. The guru claps his two hands, and then asks his disciple to concentrate on the sound of one of them. It is said that if one has grasped the meaning of this phrase he has obtained enlightenment, and is no longer under the bondage of such notions as 'I' and 'mine'.

To most Europeans, and perhaps to most Indians, this must appear to be nonsense. In my opinion it is not nonsense, but good modern physics. I only disagree with the Zen teachers in thinking that it does not necessarily lead to enlightenment, though I think that it may do so. The sound made by one hand is not an observable, that is to say it is not an object of any of our senses nor can it be detected or measured by any instrument. Nor are the psi waves which, according to some attempts to put a mathematical theory into words, are in some sense the basis of material events and objects.

Let us see how scientists are led to think in terms of unobservables. I begin with a very simple observation. If you go out before dawn you will see very little colour. As the light increases you will see green leaves and blue or yellow flowers for some time before you see red flowers. You have the same restricted colour sense as the kind of colour blind men called protanopes. Colour is so obviously dependent on the conditions of lighting and those of the eye and brain of the beholder that some European philosophers called it a 'secondary quality', as compared with 'primary qualities' like size and shape and 'tertiary qualities' like dissonance, which depend in part on education, for Indians find some sound sequences melodious which are not generally so regarded in Europe. From Newton to Planck physicists tried to explain secondary qualities in terms of primary ones, and they had some success. A red object can be defined as one which when light is shone on it, absorbs light of certain wave lengths more readily than that of others.

The attempt to explain colour, warmth, sound, and so on in terms of atomic structures and movements is part of the search for invariants, that is to say properties which are unaffected by changes in the observer or his language. For example the difference between two weights is not the same if we measure them in seers or in tolas, but their ratio is unaltered. The numbers of atoms of various kinds, namely hydrogen, oxygen, sodium, and chlorine are not altered when a lump of salt is dissolved in water, but their pattern is altered. And we may say that it has altered because the observer has got older.

The notion of an invariant was first made precise in mathematics. We can write down the equation for a curve such as an ellipse. We can then ask what algebraical expression remains constant if the position of the ellipse is altered provided its shape and size remain the same. This is not a very difficult problem. It is a little more difficult to answer the question of what is invariant in the set of all paths, in space and time simultaneously, possible for a lighter body moving round the sun, if we give a rough solution ignoring the pull of the planets. It is very much harder to answer this question if we allow for the gravitational pull of the planets.

From a mathematical point of view the search for 'laws of nature' is a search for invariants. A big step towards invariance was made when Einstein formulated the special, or restricted, theory of relativity, which allows us to describe large-scale events in a way which is invariant if an observer is moving at any constant velocity. The general theory, which holds when an observer is changing his velocity, is still incomplete. The difficulties of describing small-scale events are much greater, because the act of observing a system changes it. We can only see an object if it alters the course of some ray of light from what it would have been were the object not there, and much the same is true for the other senses. If the object is small enough the light will alter its state of motion, and we cannot simultaneously measure where it is and how quickly it is moving. Because the act of observation alters a material system, any objective account of it, that is to say any account which is the same for everybody, whether or not they are observing the system, must be in terms of unobservable events or objects. This does not mean that materialism has broken down. The opposite is true. Two conscious beings cannot experience the same small-scale event. So any attempt to describe the world in terms of experience common to several conscious beings is bound to be inaccurate. Idealism will not work on the atomic scale. But the situation for materialists is not simple either. In Lenin's words, or rather those of his translator 'The sole property of matter, with the recognition of which materialism is vitally concerned, is the property of being objective reality, of existing outside of our cognition.' The best account of objective reality, in this sense, which mathematical physicists can give us is in terms of the probability of a particle being in a given small volume during a given short time. This is a long way from the mechanistic materialism of textbooks of elementary physics. It is perfectly compatible with Lenin's view when he wrote 'It is of course totally absurd that materialism should maintain the 'lesser' reality of consciousness or should necessarily adhere to a mechanistic world-picture of matter in motion and not an electromagnetic one, or even some immeasurably more complicated one.'

If I wanted to develop the analogy of the two hands in the Koan, I might try to explain how a pair of conjugate complex quantities, as mathematicians call them, occur in modern descriptions of 'objective reality'. But it may be more illuminating to describe the actual history of the discovery of the particles called mesons. When electrons are surging backwards and forwards in a metal rod they generate radio waves, which in turn can make electrons move to and fro in a radio receiving antenna. The electrons in an atom are also believed to be moving quickly, but they do not generate observable radio waves or light. Yet the fact that they are attracted to the positively charged atomic nucleus can be explained if they are exchanging flashes of light, or photons as they are called with this nucleus. These photons are unobservable. If they exist they are private light, so to speak. An atomic nucleus consists of a number of particles, some with no electric charge, some with a positive charge. It would fly apart unless there were some very powerful force holding it together.

Yukawa, a Japanese mathematical physicist, showed that the particles would hold together if they were exchanging particles whose mass he calculated roughly. He did not, so far as I know, suggest that they could be observed. It would be most interesting to know whether Yukawa was directly influenced by Zen Buddhism. Of course no cultured Japanese can avoid being influenced by it indirectly. Some years after Yukawa's publication a study of so called cosmic rays, more accurately very quickly moving particles from outer space which are constantly entering the air, showed that after collision with atoms in the air, they generate particles whose properties agree with those postulated by Yukawa. They are now called mesons from a Greek word meaning middle or between, because their masses are between those of an electron and a nucleon, that is to say one of the components of an atomic nucleus. There are several kinds of mesons, and the kind which Yukawa imagined only lasts for about a fifty-millionth of a second on average, when it turns into a somewhat lighter and stabler kind. However, it makes its existence known in other ways than by holding atomic nuclei together.

This is quite typical of how physics progresses. Objects are postulated to explain some observed fact. Thus atoms were postulated to explain why, for example, the weight of oxygen per unit weight of carbon is twice as great in one of the more easily prepared oxides of carbon as in the other. But it was nearly a century after Dalton put his atomic theory forward that individual atoms were seen by the flashes which they make when moving very fast. Of course a good many of the objects which physicists have imagined, and which have helped them in their work, such as tubes of force and electronic orbits, have not made good their claim to reality. I do not think that some more modern conceptions, such as molecular orbitals, have much of a future, useful as they are at the moment.

Let me end up with one more example. A Zen guru might ask a pupil to contemplate the wings of a caterpillar. A caterpillar has no organs of flight. But it has what are called imaginal discs which are groups of cells which will develop into wings if the caterpillar lives. I have little doubt that when we know more we shall be able to make a caterpillar develop wings without altering its other structures. Professor Maheshwari and his pupils in the Department of Botany at Delhi are already doing something of the kind with plants.

To come back to the sound of one hand, sound is a form of energy, and each hand possesses energy of motion just before they meet. When they meet, some of this energy appears as sound. The better we understand the transformations of energy, the nearer we are to enlightenment, for the more ludicrous do our 'common-sense' notions of 'I' and 'mine' become. I know that it is not fashionable to suppose that the study of material happenings can lead to enlightenment and liberation. We are enjoined to study science to improve the means of production. This is no doubt one reason for doing so. Perhaps it is not the most important.

44

Hitting the moon

Mr Selwyn Lloyd, the British Foreign Secretary, is reported to have said, when informed that a missile launched in the Soviet Union had reached the moon, 'I don't think many people are terribly interested in it.' This is as appalling a comment as could be imagined on the kind of people Mr Lloyd meets. I hope it will lose his party a million votes in the coming election. For the people of Britain ought, quite literally, to be 'terribly interested'. For if a missile cannot merely hit the moon, but a region of it determined within 200 miles, as has been stated, it can probably hit a target in England with an error of one mile. Curiously enough this probably makes the people of London a little safer. In the last war the Soviet bombs, unlike the British and American ones, were mostly dropped on targets of military importance. And in a war with Britain the first few hydrogen bombs would probably be directed at rocket launching sites and aerodromes rather than cities. Some are so near to London that a bomb burst over them would cause very heavy damage and casualties in London, but would probably leave most of its inhabitants alive, and possibly somewhat critical of their government's foreign policy.

Mr Eisenhower, who has commanded a great army, knows that wars are now largely decided by technical equipment. So I have no doubt that he is interested. But Mr Lloyd apparently thinks that he is Lord Palmerston or Sir Edward Grey. He is not. People are no longer 'terribly interested' in British foreign secretaries, because successive British governments have thought other matters more important than scientific teaching or research. I was in Malaya recently, and saw the colossal effort devoted by the British government to hunting down a few hundred communist rebels. These men and women have now mostly been killed or captured. But the money spent on liquidating them could have trained some ten thousand engineers, who are considerably more of an asset in any future war than Singapore was in the last war or is likely to be in any other. In other words this small group, even if they have personally been defeated, have won a major victory for communism.

Mr Lloyd and his acquaintances are no doubt genuinely uninterested in nature, and cannot understand why other people are so, except perhaps in so far as money can be made out of it. Even if the Soviet missile had done no more than reach the moon, it would have made a notable contribution to science. If the report is true that a Hungarian astronomer has seen a dark circle spreading out on the moon from the point of impact and lasting for an hour it tells us a very great deal. Many astronomers think that the relatively flat parts of the moon, called Maria, or seas, which they certainly are not, are plains covered with dust from the disintegration of rocks by the alternation of heat and cold, and the impact of meteorites. Now a projectile hitting a plain covered with dust would doubtless raise a dust cloud; and on earth such a cloud might remain for some hours. This is because the dust particles are raised very quickly, but fall very slowly through the air. But if there is any gas round the moon, its density is probably less than a millionth of our own. Particles of dust too small to be seen would therefore fall back as quickly as large stones. I only know of one agency which would keep a cloud of dust suspended above the moon's surface for an hour, in the absence of air. That is electric charge. A cloud of particles all with the same electric charge would stay up for a long time, because each would repel the others. Only the particles on the surface of the moon can be charged, but once they are thrown upwards they may acquire a charge very rapidly under the influence of the X-rays and ultraviolet radiation from the sun, which are stopped by the upper layers of our own air, and produce the charged regions which reflect radio signals.

It has been calculated that a meteorite displaces about sixty thousand times its own weight of material. As the lunik weighed 400 kilograms, it would displace about 24,000 tons of lunar material, and if even a third of this went up, it would form a considerable cloud. It may be possible to photograph the crater formed by the missile. This can best be done when the sun is nearly setting or just risen in the area of the moon concerned, for then shadows are long. In other words the area of impact should be observed when it is in the bright part of the moon, but near to the dark part. Perhaps the next good opportunity may be when the moon is waning. Anyhow we shall get some information about the moon's surface, including perhaps valuable data as to the possibility of making a relatively gentle landing on one of the seas of dust.

But it is more certain that we shall get a lot of information about the space between the earth and the moon. Has the moon got a magnetic field like the earth and the sun, not to mention a few stars whose magnetism is far more powerful than the sun's? We now know that its magnetic field is far smaller than the earth's. Nobody knows the cause of the earth's magnetism, but it may be due to electric currents eddying in its metallic core. The moon is much less dense than the earth, its average density being just 60 per cent of the earth's; and it probably has no metallic core. If the moon is a magnet at all comparable with the earth we shall have to scrap the theories about the magnetism of the earth and the stars which are at present most popular.

We shall also get a good deal more information about the charged particles which have been caught in the earth's magnetic field and are moving in various orbits with very high speeds. The existence of regions in which such particles are moving was predicted by Stormer, a Norwegian physicist, in 1911. However, his calculations were wrong in detail, though Chapman in England later predicted the correct distance from the earth, and American popularizers write of 'von Allen radiation' after an American worker who deduced their existence from the signals of American satellites. Whether they were first discovered in the USA or the USSR I do not know. There is a substantial agreement about their existence and properties. Human beings who are shot up far above the air will need protection against them. But their study is important for other reasons.

Much of our knowledge of the fine structure of matter is derived from the study of what, in my childhood, were called vacuum tubes, but which in fact contain gas at densities which are low by the standards of the earth, but very much higher than in the space between us and the moon. In such tubes electrically charged atoms and molecules can move for fairly long distances without collisions, and display properties which were previously unsuspected. Nowadays we can make them move in nearly circular orbits in cyclotrons and similar expensive apparatus, where the path length may be some kilometres. But in the space between the stars they can travel for years without collision. In these circumstances they may be expected to show new kinds of behaviour, as men do when they isolate themselves from contact with their fellows.

In the last twenty years a new branch of physics called magneto-hydrodynamics has developed. It treats of the motions of charged particles in magnetic fields, and may be going to explain a great deal about the observed structure of stellar systems. It is not only difficult mathematically. It is as yet based on regrettably few observations and still fewer experiments. The 'sputniks and luniks are signalling back results which will play an important part in the development of this science. The American propaganda service claims that their satellites are provided with apparatus so much better designed than those in the Soviet satellites that they furnish more information, though lighter. This may or may not be true. What is certain is that much of the information on these matters from the Soviet Union which reaches the Indian press is grossly distorted, perhaps in translation, and the Soviet authorities concerned could do worse than improve the information service from Moscow to India, however good that from Bhuvaloka to Moscow may be.

There is perhaps an even simpler reason why it is scientifically important to have hit the moon. It is a concrete proof that certain physical 'laws' or rules about the behaviour of matter hold good at least as far as the moon. Some of these rules break down at very small distances, others at very large ones. Others may be superseded by unexpected properties of matter at intermediate distances. The fact that we can plan and carry out a journey to the moon, so far only by machines and not by men, makes it much more likely, but not certain, that we could send similar missiles to Sukra or Bṛihaspati without taking hitherto unknown agencies into account. If we had got that far, it would not prove that we could send them to distant stars.

Some Indian politicians have spoken rather flippantly of this work. I think they may not realize its scientific importance. Personally I think that biological research is even more important both practically and theoretically. The next few months will show whether I am right or wrong in my beliefs as to what kind of research is practically important. But biology is not so useful as physics either for making profits or for killing people. So it has not got the same prestige in the present human society.*

45

The dog in the sputnik

It is now officially announced that the dog in the sputnik is dead. Either there was no way of sending her down, or the device for doing so did not work. English dog-lovers protested to the Soviet embassy. I have yet to learn that English bird-lovers protested to Mr Macmillan against exploding thermonuclear bombs over the sea, where they must kill many thousands of sea birds outright, and many thousands more in a slow and painful manner.

I cannot blame the people who sent Laika to her death for a very simple reason. I should have been quite ready to take her place.

* Haldane did not live long enough to see the golden age of molecular biology when the 'prestige' of biology matched or exceeded that of physics, although he predicted that possibility in his book: *Daedalus or Science and the Future* (1923).

Though in fact I am much too heavy for the job. If a human volunteer is sent up in another sputnik with a small chance of coming down again alive, or with none, it should be someone weighing less than myself, and preferably with incurable cancer or some similar disease, but not too ill to report back until they die.

I do not know how the dog Laika felt

> 'When her virginal feet were set
> On the terrible heavenly way.'

Dogs usually enjoy experiments. I have seen one in a laboratory at Leningrad pulling a human assistant along a corridor in its eagerness to act in an experiment of the Pavlov type on conditioned reflexes. There is no reason to think Laika felt any great fear or pain at any stage.

I probably know better than anyone else in India how a human volunteer in a similar position would feel. Of course I have never chosen certain death, but I have often chosen a fair risk of death. This produces a feeling of exaltation which makes it hard to give an objective account of what is happening. A soldier's account of his first battle is probably quite unreliable.

The first possible danger to a passenger arises from acceleration. If a man is being pushed violently his blood tends to remain behind. If he is being accelerated hand first, his blood tends to collect in his feet, and he may faint. If he is accelerated feet first it is forced into his head, and a blood vessel in his brain may burst. If, however, he lies down he can stand an acceleration of five or six times that due to gravity for a minute or so. All this has been worked out both in aeroplanes and in centrifugal swings.

After a few minutes at most of violent acceleration the sputnik is launched in its orbit, and is in a state of 'free fall', the earth's pull on it being balanced by the centrifugal force with which it is 'Struggling fierce towards heaven's free wilderness'. (I apologize for possible misquotations from Swinburne and Shelley. My books have not yet arrived from London.) The passenger has no weight. There is no up and down for him or her. Men have experienced this state for times up to about 20 seconds in diving aeroplanes. Some were upset by it, some enjoyed it. But even the latter might develop symptoms like sea sickness after a few hours. If Laika did not, this is a valuable bit of knowledge.

The nearest condition to weightlessness which I have experienced is life in a flexible diving suit in which one's head is enclosed in a waterproof bag, not a metal helmet. The tendency of the blood to run into ones head, which is so unpleasant when one hangs head downwards in air, is completely counterbalanced by the increased pressure of the water. One of my colleagues could go to sleep under water with his head downwards. I never did, but my last dive was in a cave where I had to go down a narrow vertical shaft from which water was rising, for about 15 feet head first, before starting to crawl along the channel through which it came. This is no harder than riding a bicycle, but of course it needs some practice.

In such a dress one still knows which direction is upwards, from small organs inside the ear containing heavy particles. But this knowledge makes little difference to ones feelings or actions. And a diver soon learns to screw up the valve through which air or oxygen escapes until his net weight is only five or ten pounds, and another turn of the screw would inflate his dress so that he floated up. I would recommend any intending passenger in a sputnik to spend forty-eight hours or so lounging under water in a flexible self-contained diving dress.

Laika was very possibly supplied with food and drink by injections. In the absence of gravity one cannot pour water from a bottle. A human being should probably take it from a bag with a porous mouthpiece. However breathing is much more of a problem. Here too I can speak with some personal experience. I have been shut up in a metal box with no means of renewing or purifying the air till I was nearly unconscious, and have often lost consciousness from lack of oxygen or excess of carbon dioxide. I have also been shut up with a companion for forty-eight hours in a metal chamber which was in fact a compartment of one of the miniature submarines used by the British navy during the second world war. (Many writers write war with a capital letter. I regard this as the sign of a devotion to Skandha* which I do not feel.) We had adequate supplies of oxygen and of an absorbent for carbon dioxide, and suffered from nothing worse than cold, wet, and stiffness. For the compartment was not very large, and we could not both stretch our legs at the same time.

If one is on the way to dying from lack of oxygen, one pants, but not very violently. If carbon dioxide is also accumulating, as it does when one is shut up without any absorbent, one pants very heavily. But there is no pain, and one gradually falls asleep. I have also been buried alive under sand, and had to breathe through very narrow tubes or cotton wool pads. This is most unpleasant. I felt as if my chest would burst, and in fact partial obstruction of the breath has been used as a torture. But provided one can breathe freely, the panting is no worse than after running a mile or so. I would far sooner die from exhaustion of the air in a confined space than from most diseases.

When black mice were sent up in American balloons with adequate oxygen supply to a height of 20 miles they developed some white hairs at their next moult. They had been hit by atomic nuclei moving with very nearly the speed of light which went into their tissues and killed all cells along their tracks, including some of the cells making black pigment. These atomic nuclei came from outer space, and are stopped by the relatively thick air of the lowest 20 miles. Their properties have been studied from the tracks made when they penetrated piles of special photographic plates. I don't think such particles could kill a dog in a week. They might reduce a man to stupidity in some months, and impair his vision. Space ships may have to be built with very thick metal walls to keep them out.

Why have the Soviet Union achieved this wonderful result when the United States are well behind them, and other governments have not even tried? No doubt the Soviet system is specially favourable

* God of war in Hindu mythology.

to large-scale engineering, but the Vedas give another reason. 'Devas tam parāduh yo nyatrātmano devān veda', the gods desert him who knows the gods non-spiritually, I read in the Bṛihadaranyaka Upanisad. The Americans are constantly repeating a formula stating that God is in heaven, and it must influence their thought even if they do not believe it. They might do better to say that heaven was in God, though this too seems to me a crudity. If you think of the sky as God's special abode, you are likely to be inefficient in dealing with it. Perhaps I am old-fashioned in thinking that mantras are important. But this seems to me to be a case where they have had some historical influence.

46

What I want to know about Gagarin

Yuri Gagarin is perhaps the most famous man in the world today. And it is good that such a man should be famous. For millions of boys all over the world must want to tread '*On the terrible heavenly way*' which he has opened, and if they cannot be the first to get above the air, or even to reach the moon, may hope to be the first to land on Mars or on an asteroid. And it is easier to try to imitate Gagarin's good qualities than those of the mathematicians and others who made his achievement possible.

It is obvious that Gagarin is a brave man. But there are several types of brave men. To judge from his appearance, which may be very misleading, and his few recorded sayings, he is of the type which, when exposed to danger or pain, behaves with complete calm. He probably did what had to be done almost exactly as he had done it in practice at ground level. There is another type of brave man or woman who is often rather unreliable in normal life, being prone to accidents, encounters with the police, and so on, but is at his or her best in a dangerous or at any rate frightening situation. I chose such people as colleagues in underwater work twenty years ago.

Unforeseen

Perhaps the extreme example of such a man in literature is Dolohov in Tolstoi's *War and Peace.* I wanted these qualities because we were working with makeshift machinery, and all sorts of unforeseen things went wrong. For example, the first time I got out of a miniature submarine under water, I had practised opening the lid of the escape chamber in air, but when it was flooded under water there was a lot of oil on top of the water. I was wearing an escape apparatus, and when the water rose over my head and I tried to turn the wheel to open the lid, I found it covered with oil and my hands slipped. I was quite frightened, as I had no means of communicating with others. I finally got the lid open and recommended changes in the design including a milled edge to the wheel.

I have no doubt that all the apparatus which Gagarin had to use had been tested many times under all kinds of stresses, and the various ways in which it might go wrong were pretty well known. If so, the calm type of courage was needed. But when the first human party lands on the moon they may be exposed to unforeseen dangers, and I hope they will include at least one man with the good qualities of Dolohov, or at least of Denisov.

Gagarin is also no doubt an intelligent man. But I am sure he is not a man whose intellectual processes interfere with his practical activities. I hope that one of the next astronauts will be a poet, but if so he must have an unimaginative companion who will turn the necessary handles while the poet is trying to find words for experiences which no human being has had before.

Acceleration

Gagarin was also no doubt chosen for his physiological capacities. When going up and coming down he was subjected to intense accelaration. He lay with his body at right angles to the direction of acceleration, or most of his blood would have been forced into his head or feet. But even so the strain may have been considerable. I have no doubt that a large number of people were tested for resistance to acceleration, and that Gagarin was one of the most successful 10 per cent.

I have also no doubt that he is a good all-round athlete. He was also, I think, chosen partly because of his small weight. Had I been in charge of the choice, I might have recommended a woman on this ground. Courage of the Gagarin type is quite common in women, though most frequently evoked by threats to their children.

For more than an hour he was subjected to a condition which no one had experienced before for more than a few seconds, namely weightlessness. Even if his craft was spinning quickly enough to force him gently to one side, he must have felt as if he were falling. Now when most people are subjected to abnormal episodes of falling and rising they get seasick. It is possible that normal people would be in this distressing state after half an hour's 'free fall'.

I have found that I can usually prevent seasickness or air-sickness for an hour or so by controlled breathing. I breathe regularly and deeply at intervals of ten seconds or so and this holds off vomiting. But it is quite possible that Gagarin was chosen because, among other things, he was immune to seasickness. If so, it is very important to find out whether in less well-qualified people weightlessness leads to vomiting or any other symptoms.

I do not doubt that Gagarin was tested by exposure to various kinds of stress. Heat and cold were doubtless included, for a spacecraft is bound to be heated while moving very quickly through the air on its way up and down. And while circling round the earth it will pass about half an hour in intense sunlight, and almost as long in intense cold.

Provision is, of course, made to keep its temperature steady, but it might go wrong. One's reactions to heat and cold depend largely on the control of the heart and blood vessels. They are not directly under conscious control, but the mind certainly influences them.

Similarly he was no doubt tested with psychological shocks such as loud sounds, flashes of light, and so on, to see that they neither affected his heart nor stopped him doing a job. In both these sets of tests he was probably helped by joining the Communist Party, as he has recently done. The acceptance of Marxism leads to the disappearance of certain kinds of worry, which may play a part in causing gastritis, anomalous heart beats, and so on. And membership of the party might suffice to give an extra few minutes' endurance of extreme hardship.

Experiment

If I had been in charge of his training, I should have tried another experiment on him. I would have shut him up with no supply of oxygen or means of absorbing carbon dioxide, until he lost consciousness. If carbon dioxide is removed, one loses consciousness gradually without any discomfort. But if carbon dioxide accumulates, one pants very hard indeed for some time. This is not painful, but can be alarming. Gagarin's reactions to these stresses would have been noted.

If the arrangements for returning him from his orbit had broken down, he would have died from want of oxygen or excess of carbon dioxide. And, as he is a brave man, the knowledge that it is a painless form of death would have helped him to go on reporting calmly to earth if he believed that he would not come down again alive. Fortunately these tests, even if they were made, were unnecessary.

Above the atmosphere there is a variety of radiations and of quickly moving particles which are stopped more or less completely by the air. If we lived on the surface of the moon we should doubtless be resistant to them. But, in fact, they are harmful to us. Some of these rays and particles, such as X-rays from the sun, and electrons discharged from the sun or moving round the earth, are fairly completely stopped not only by a spacecraft's metal skin, but by its window which lets light through. However, some particles, and particularly the nuclei of atoms heavier than hydrogen, are moving so fast that they can not only penetrate metal plates, but in doing so produce showers of other particles such as mesons.

No doubt Gagarin was partly protected against these particles, which are misnamed cosmic rays. But unless discoveries of an unexpected kind have been made about them in the Soviet Union, and kept secret, which I greatly doubt, the protection was not complete.

I think that in the next few months Gagarin is likely to develop a few patches with white hairs where a rapid particle has killed the pigment-forming cells. These have been seen in mice which have been for some hours at great heights. I have little doubt that physicians are examining Gagarin's skin for other possible effects. They are not likely to be immediately serious.

Blind Patches

Again, small blind patches may have appeared in his retinae, the light-sensitive membranes at the back of his eyes. They are not likely to spread, and are quite harmless, as we all have large 'blind spots' already, which we do not normally notice. But if he has got even a few blind spots in one hour in space, there may be a danger of total blindness in a three years' voyage to Jupiter or Saturn.

He is probably a little more likely than before he went up to develop certain kinds of cancer. Space travel may be as dangerous as cigarette smoking from this point of view. And high energy particles seem to have a general aging effect. He may have diminished expectation of life by a week. This is not a serious matter, but the effect of a year's voyage in space might be very serious indeed. So the medical examinations to which he will be subjected for the rest of his life may be of great importance to future space travellers, even if the dogs, which were up for longer yield more information.

I make these simple remarks because I have little doubt that some people will say that the results of this wonderful feat of technology and courage can be of no scientific importance. On the contrary, they may be of value in such unexpected quarters as the studies of aging and of blindness. The knowledge that a man can behave normally when weightless, and immediately after being so, is most important, and I must confess that I had considered it possible that the first cosmonaut would be seriously affected by this condition.

It is not a chance coincidence that the combination of intelligence, technical ability, and courage needed was first realized in the Soviet Union. One of the lessons for India is this. Our educational system is modelled on that of Britain. It is unsatisfactory in many ways, and the need for reform is generally admitted. Is it not time that those who are concerned with this reform should very seriously consider copying many features of the educational system of the Soviet Union which is certainly leading the world in many branches of science?

47

Some autumn stars

Most educated people have at least a rough idea of the sun, moon, and planets, and their movements. They know very little about the background or 'fixed' stars, except that they shine by their own light, like the sun, but unlike the moon and the planets, and that they are very far away. It is only in our lifetime that we have got to know much about them, and I know of no book which gives even roughly up-to-date information about individual stars, though several give a general account. I am therefore going to write a series of articles about a few of the more conspicuous stars to be seen at various times of the year. I am handicapped for two reasons. I cannot ask for the publication of a star map, and I do not know the Sanskrit names of most of the stars, and only those of a few of the constellations. In Europe and North America most of the names of individual stars are Arabic, while those of the constellations are Latin or Greek. Perhaps a reader will help me with a good reference to Indian astronomical names.

What do we know about a star ? If it is bright enough to have a name, it is always near enough for us to measure the distance. This is a fairly simple matter. If we photograph the same bright star at an interval of six months, say in March and September, it will appear to have moved slightly relative to the fainter and farther stars in its neighbourhood. This is because the earth has moved relative to the sun. Half of the star's apparent movement is called its parallax. It never reaches a second of arc, that is to say the angle subtended by a centimetre kilometres away. However, it is possible to measure about a thousandth of this angle. The principle involved is the simple one which accounts for the apparent movement of a near object against the background when one shuts first one eye and then the other. However, the base line is not the small distance between human eyes, but that between the earth's position in March and September, namely 15 crore kilometres. The unit of distance is a parsec, at which the parallax of a star would be a second. This is 30 million kilometres, and it takes light three and a quarter years to travel so far. We can easily measure the brightness of a star, compared with another star, or a standard lamp at a known distance. We can also measure the heat reaching us from the brighter stars. So if we know the distance and brightness we can say that this star gives a hundred times as much light as the sun, and that other only a tenth. We can measure the temperature of the outside of a star by two methods. Cool stars are red, and hot stars blue. But we can do it a good deal more accurately by a method due to Meghnad Saha which depends on the star's spectrum. This enables us to calculate the star's size if its luminosity is known. In the case of double stars revolving round a common centre of gravity we can also calculate the mass. Finally we can measure the rate at which a star is approaching or receding, from its spectrum, and if it is near enough we can often measure how fast it is moving sideways.

Now let us consider a few of the stars visible in India in September evenings. Two bright stars are not far apart in the south, and set before midnight. These are the planet Saturn or Sone,* and the brightest star in the constellation Scorpius or Brischik, which is called Yestha or Antares. Antares is a red double star, or more accurately a pair of stars of which one at least is about 20 times as massive as the sun, and of about 400 times its radius. However, its temperature is only about 3,000 degrees, or half that of the sun. But it is so large that it puts out about 2,000 times as much light per second as the sun. So though it is about 120 parsecs away, it is the fifteenth brightest star in the sky.

As we glance northwards along the Milky Way, the next very bright star to be seen is Altair, in the constellation of the Eagle. This is one of our nearest neighbours, just under five parsecs away. It is about 20 times as bright as the sun, not quite twice as heavy, and about one and a half times as hot. With the sun, it belongs to what is called the main sequence of stars. It appears probable that all stars start as members of the main sequence, in which there is a simple relation between mass and light output. So long as their interiors contain enough hydrogen their light output alters little for a time which may be measured in thousands of millions of years. Altair is a much more representative star than most of the bright ones.

As our gaze moves further northwards along the Milky Way we see Deneb, the brightest star in the constellation of the Swan. It is so far

* This was true in 1958.

away that its distance is uncertain, though probably about 500 parsecs. It has by far the greatest light output of all the stars except Regel in Orion which is easily visible with the naked eye, about 50,000 times that of the sun. Though it is only about twice as hot as the sun it has 4,000 times its area. Such stars are very rare, only nine being visible, though they could be seen at a distance of 5,000 parsecs, whereas the sun could not be seen at more than 30 parsecs.

Vega, in the constellation of the Lyre, not very far west from Deneb, is a main sequence star about eight parsecs away and not unlike Altair, though giving out about twice as much light. The earth's axis gradually spins round, and although it now points to Polaris, or Dhruva, it pointed approximately to Vega 13,000 years ago, and will do so 13,000 years hence. Our present pole star is about one and a quarter degrees from the pole, so that it appears to go round in a very small circle. It is a double star, and about 120 parsecs away, like Antares. However, the two stars which compose it are too near to be seen separately with an ordinary telescope. They are detected as follows. If a star is moving towards the earth the light characteristic of any particular change in the electrons surrounding an atom shifts towards the violet, if it is moving away, towards the red. Such shifts occur in the spectrum of Polaris with a period of about thirty years. This is almost the same as the period of Saturn's revolution round the sun. However, Polaris, though not much hotter than the sun, is ten times heavier, so the star revolving round it must be a bit further from it than Saturn is from the sun. Sirius, the brightest star, has a faint companion which goes round it every five years, and was first detected spectroscopically, and only later seen with a telescope. Probably the companion of Polaris will be detected directly in the future.

I am sure that this article has annoyed some readers. It contains too many figures. On the other hand the scientific, as opposed to the mathematical, point of view, requires some knowledge of familiar objects. The stars are among the most familiar, but what we know about them is somewhat alien from ordinary experience.

Nevertheless, Indian scientists, notably Saha and Chandrasekhar, have made contributions to astronomy much more important than any Indian contributions to chemistry, for example. I have therefore ventured to put a few results of modern astronomy before my readers, and hope to write three or four such articles a year, so that in the course of a year or two I shall have told them something about the thirty or so stars which, either because they are near us, or because they emit a great deal of light, appear brightest to us dwellers on earth.

48

The Pleiades and Orion

Perhaps the most striking of the star groups to be seen in late autumn is the Pleiades or Krittika. Most people can see six, perhaps a few seven, without optical aid. Even a modest field glass shows many more. But the counting of them is an extremely difficult task because there are so many other stars in the same direction. In fact the Pleiades are a natural group, that is to say they are not merely in the same direction but close together. So all the stars which belong to the group must agree in four measurable characters. Firstly, because their distances are nearly the same, they must agree in parallax, that is to say in their apparent movement against the background of faint stars beyond them as the earth moves round the sun each year. The distance is in fact about 130 parsecs, or 420 light-years. The light which reaches our eyes today started off about AD 1540 say about the time of Babur's invasion of India.

Secondly they must all appear, relative to their background, to be moving in the same direction with the same speed. In fact they appear to be moving to the south-east, towards a point in the southern constellation Argo. When allowance is made for the sun's movement relative to its neighbours the point is somewhat further from their present direction. A simple analogy will make the matter clear. If we watch a flock of birds flying together in the same direction, they appear to be converging towards the same point on the horizon. To a man in a moving train they would seem to be moving to a different point. Finally all these stars are moving away from us at about the same speed of 20 kilometres per second. If a star agreed with the rest in only three out of four of these measurements, it might be in the neighbourhood of the Pleiades at present, but it would have been some distance from them 100,000 years ago.

There are about 120 stars which, on these tests, are members of the Pleiades, and in addition there is a good deal of nebular matter—shining gas and dust—between them. There is plenty of evidence that such matter is associated with recently formed stars, and the Pleiades are probably less than 10 million years old. It is generally thought that stars are formed from the condensation of such clouds. However, at a conference held at Lucknow in 1955, of which a report has just been published by the National Academy of Sciences, India, Ambarzumian, an Armenian who is one of the leading Soviet astronomers, questioned this theory. He thinks that the nebulae may be formed by the same unknown process which also provides new stars.

However that may be, the Pleiades have a wide range of luminosities. The visible ones are much more luminous than the sun, but the dimmest members so far identified are less so. And probably when still dimmer stars in this direction have been investigated a thousand or more Pleiades will be found. But as there are plenty of dim stars in their direction, new members can only be identified after careful measurements. The stars in such a cluster must be moving relative to each other, or they would fall together. If any moves too fast, it will be able to escape from the gravitational pull of the others. And this must occasionally happen when one of them acquires enough speed by passing close to several others. Such an event may occur at intervals of a few hundred thousand years, and Chandrasekhar has calculated the probable 'life' of an open cluster like the Pleiades. This is certainly a good many million years, but quite a short time compared with our earth's age.

Rohini, or Aldebaran, rises soon after the Pleiades, though at present [1958] it is less conspicuous than Mangal, or Mars, in much the same direction. It is a double star about 21 parsecs away, red in colour and thus cooler than the sun, and a good deal more luminous.

Then follows Kalpurush, or Orion. This is not a 'natural' system like the Pleiades. All the bright stars in it are distant, but some are twice as far away as others. However, it includes some clusters. The three 'stars' which form the belt are at a distance of about 200 parsecs. If they were as near as Sirius they would appear so bright as to be visible by daylight, and we should see with the naked eye that Mintaka, which rises first, was not a star, but a system of at least three stars. One is so far from the other two that there would be no difficulty in seeing them apart. However, two are very close together, spinning round their common centre of gravity in a little under six days. These are so close that they cannot be seen apart with the best available telescopes. However, the spectrum of a star gets bluer as it approaches us, and redder as it moves away; and a large number of apparently single stars have been shown, in this way, to be double. In some cases a more powerful telescope has later enabled astronomers to see or photograph them as a pair.

The second star in the 'belt', Alnilam, appears to be single, but the apparent speed at which it is moving away from the earth varies. So it probably has a faint companion. The third star Alnitak is a triple system much like Mintaka. All three are extremely hot stars emitting about ten thousand times as much light as the sun from only about a hundred times its area. The other four brightest stars in Orion are all very distant and luminous. The brightest, Regel, at the southern end, is a quadruple star. The second brightest, Betelgeuse, the red star at the northern end, is a huge object which, if it replaced our own sun, would reach nearly halfway to the earth. It was the first star whose diameter was actually measured, not directly, but by using the principle of interference. Its light varies not very regularly over a period of about six years.

Finally in the 'sword' of Orion, between the belt and the last bright star to rise, there is the famous nebula surrounding some of the fainter stars. This can just be seen with the naked eye, but is a truly magnificent spectacle when seen through an ordinary telescope. It is 400 parsecs away, further than any of the seven brightest stars of the constellation. It is a cloud of gas of extremely low density, far lower than anything obtained in a vacuum tube on earth. However, it shines much like the gas in a human low pressure discharge tube, the atoms giving out what is called a bright line spectrum, so that when it is photographed or seen through a prism there are a few spots of light of different colours, instead of a continuous rainbow-like band, as with ordinary stars. The gas atoms glow in this way because they are excited by stars embedded in the nebula which probably emit enough ultraviolet radiation to do so. Some smaller nebulae in Orion merely reflect the light of neighbouring stars from dust particles. Only a spectroscope can distinguish the two types. There is reason to think that new stars are being formed in this nebula. Several rapidly moving stars now seen against other constellations are moving in paths which trace back to this region of space.

I hope that in this article I have written nothing but the truth. However, astronomy is at present progressing so quickly that new facts have probably been discovered in the last few years which I should have incorporated had I known them. Perhaps Alnilam has been definitely shown to be double. Perhaps a hundred more stars have been assigned to the Pleiades. It is not easy in India to keep up with the latest news about sciences which are not one's own profession.

49

Some neighbouring stars

Soon after Orion, or Kalpurush, two bright stars, Sirius and Procyon, which in the Greek mythology are the hounds of the hunter Orion, rise in the East. Both of them are near to us, which is why they appear so bright; and both are similar in other ways. Sirius is the brightest of the 'fixed' stars, and one of the least fixed. It is so near that its light reaches us in under ten years, and its apparent motion is over two minutes of arc per century. If an astronomer who lived two thousand years ago could be brought to life he would not notice any change in the apparent shape of Orion, though these changes can be measured with modern instruments. But he would notice that both Sirius and Procyon had moved through angles of over half the moon's apparent diameter. The only other bright stars which have changed their direction more than Sirius are Alpha Centari, which is nearer, and Arcturus, which is moving very rapidly.

Sirius and Procyon are both more luminous than the sun, that is to say they give out more light, but only about twelve and seven times as much respectively. They are also about twice as heavy. We know their masses within 1 per cent or so, because each has a very faint companion, and the pairs revolve round their common centres of gravity with periods of fifty and forty-one years, in elliptical orbits which are much less circular than those of the planets in our solar system. So we can calculate the masses of both stars by the same methods which enable us to calculate that of the sun from its gravitational pull on the earth. These companions are very queer objects indeed. The companion of Sirius, or Sirius B, has the same mass as the sun within 1 per cent, and its surface is a good deal hotter. But it only gives out one three hundredth as much light and its radius is only about one forty fifth of the sun's. This means that its density is about ninety thousand times that of the sun. In other words a cubic inch of its material must weigh over a ton, and the density near the centre must be about a million times that of water. When these figures were first obtained thirty-five years or so ago, physicists found them incredible. But a number of other such stars have since been found and there seems little doubt of their truth. These stars have collapsed under their own gravitation, and it is thought that this is the normal fate of all stars when their sources of energy run down. They are called white dwarfs, to distinguish them from the red dwarfs, which also give out very little light, but are less massive than the sun, and a good deal less hot; they are not however surprisingly dense. The matter in white dwarfs is of course under enormous pressure, and we can only guess at its properties.

A sinuous line of stars stretches southwards from Orion's foot to end in Achernar, a bright and distant southern star. This is called the constellation Eridanus, and contains a number of our neighbours, some of which are visible, though none is bright. The most interesting is Omicron-two Eridani, a system of three or perhaps four stars, of which the brightest has only one ninth of the sun's mass, and is the lightest star whose mass is accurately known, though there is little doubt that some others are lighter. However, the range of stellar masses is quite small, from about a fifteenth of the sun's mass to 400 times. Nothing lighter can, apparently, produce enough heat to shine, while anything heavier would develop so much heat as to burst. This is a surprisingly small range, of only about six thousand fold. It may be compared with the range of sizes even within a single group of animals such as the mammals, where a whale weighs as much as many million mice. The reason for the small range is no doubt that stars are atomic reactors, and the working conditions of an atomic reactor are much more stringent than those of a steam engine, a water mill, or a fish.

The nearest star which we can see which is comparable to the sun in brightness is Eta Cassiopeiae. The constellation Cassiopeia is the very familiar M-shaped formation in the Milky Way near the north pole star. It never sets for observers in Britain, but is only to be seen clearly in the evening from India in autumn and winter. If you look between the second and third of the five bright stars in order of rising you will see Eta. It is about five and a half parsecs away, and is a little hotter than the sun but not quite so luminous, as its mass and area are smaller. It has a companion which is a good deal lighter, cooler, and dimmer, but bright enough to be easily studied. The two revolve round one another with a period which is supposed to be 526 years, though I suspect a few years may be added or substracted when we have studied the system for another century or so.

If there are planets moving round the heavier star with a period of a year or so, then during half the year their inhabitants must see the companion star at night, giving them a good deal more light than our moon, but hardly any perceptible heat. It must move very slowly against the background of 'fixed' stars. Perhaps there are rational beings on one of these planets who live long enough to see it make a whole circle, as for example our planet Saturn, or Sone, makes one every twenty-nine and a half years. Perhaps they merely tell one another stories of people in the past who lived long enough to do so.

In the present state of astronomical theory we cannot state with any confidence whether planetary systems like our own are common or rare. But most astronomers guess that they are not very rare, and Eta Cassiopeiae is one of the ten or so visible stars from which the light reaching us tonight started in my lifetime which is like enough to the sun, and otherwise qualified, to have planets which are possible abodes of life. Among the disqualifications is a companion star close enough to disturb the orbits of the planets. This, for example, probably rules out Procyon as a likely sun round which habitable planets may revolve. Of course it may turn out that even when a planet has the correct temperature, the proper chemical conditions for life are usually lacking. It may be that it only originates under quite unusual conditions, or that only very rarely does life beseem into mind, even of the rather poor quality found in most of our species. However, unless there are unexpected difficulties, I think there must be many planets in our galaxy whose inhabitants are far superior to ourselves both intellectually and morally.

We ought, however, I suppose, to consider one other possibility. It may be a law of nature that whenever conscious beings discover the sources of energy which would enable them to journey from one planet to another, they use them to destroy themselves.

50

The seven rishis*

The constellation which in India is called the Saptarishi, in Britain the Plough or Great Bear, in America the Big Dipper, is now at its highest above the horizon in India soon after sunset. The seven main stars form a conspicuous group, and the middle five belong to an open, moving, cluster of stars. All seven are approaching us at about the same speed of 11 kilometres per second, a quite moderate speed for stars, being about a third of the earth's speed round the sun. The five middle ones also appear to be moving eastwards in nearly the same direction. They are also at about the same distance of 24 parsecs, that is to say the light from them started less than a century ago. The last of the seven, which does not belong to the cluster, is about three times as far away.

The two end stars of the seven are approaching us at about the same speed as the rest but their sideways motion is nearly in the opposite direction. All the stars concerned have moved something like a degree in the last thirty or forty thousand years. The shape of the constellation has hardly changed during human history, but it has changed appreciably since men made the first pictures and sculptures. So there is a chance that a representation of these stars, forming a shape different from its present one, may be discovered.

* Seven sages.

The constellation will continue to increase in apparent size for many thousands of years, and then get smaller again until, some millions of years hence, it will be a compact cluster of five stars like the Pleiades, finally crossing the equator to disappear near Capricorn or Makar.

About a hundred fainter stars in the same direction are moving parallel with the central five rishis. None of them are very heavy or hot, like the huge and distant stars which outline the constellation of Orion. A number of the fainter stars in the cluster are not unlike our own sun, but the five conspicuous stars are mostly double or triple systems with at least one member hotter than the sun. Arundati, by the way, is not physically attached to Zeta of the Great Bear, though this is a triple star in its own right.

The Ursa Major cluster, as it is called, is the nearest of a number of open clusters, and therefore harder to study than moderately distant ones like the Pleiades or very distant ones like Khi Persei which to the naked eye appears as a single star. For most of the stars in the constellation do not belong to the cluster but are a great deal further off, whereas, although there are some background stars behind the Pleiades, the task of picking them out is much easier.

It is generally thought that a cluster of stars is generated by the collapse of a nebula, which is a cloud of very diffuse gas and dust, under its own gravitation. If a cloud of collapsing gas is large enough, it will heat up so much as a result of the energy liberated by the fall that reactions between atomic nuclei become possible, and heat and light are then generated by the fusion of hydrogen nuclei to form heavier elements, so a star starts shining. In some nebulae condensations have been photographed, particularly by the Mexican astronomer Haro, which are thought to be the early stages of star formation.

However, Ambarzumian, the eminent Soviet astronomer, is more cautious. He admits that nebulae are generally associated with star formation, but it does not follow that stars are formed from nebulae. In the same way large factories are generally associated with extremely rich men, and it was long believed that they could only be built where rich men were able to accumulate large amounts of money. The experience of the Soviet Union, and of our own public sector, shows that this is not so.

The question may only be decided when the nebular concentrations have been watched for a few thousand years. Unfortunately no nebula is anything like as near to us as the Ursa major cluster, so powerful telescopes are needed. It may be settled by theoretical work. Or it may be settled by observations from a sputnik. For recent American work has opened up new possibilities. Some of the rockets which have been sent up above most of the air, and particularly above the layers where ultraviolet radiation is absorbed, have been equipped for photography using very short wavelengths. The results as to the sun's output of ultraviolet radiation of various wavelengths (which would have different colours if we could see them) are not unexpected. But Kuperian, Boggers, and Milligan reported utterly unexpected findings about the night sky. The photograph obtained was very crude. The resolution was only three degrees. Such a camera, using ordinary light, would barely distinguish two objects a foot apart and 20 feet away.

So the photographs parachuted back to earth only showed patches of ultraviolet emission. As expected there was a big patch in Orion, and there were other patches further south along the Milky Way, some probably due to the few very hot stars which are now visible in the south in southern India about sunset.

But there were also three regions far from the Milky Way, and not associated with any stars previously thought to emit ultraviolet radiation. One is in the Lion, one near the last of the seven rishis, and the largest of all round Chitra or Spica, the brightest star in Kanya or Virgo. This has a diameter of about 22 degrees, one quarter of the angle between zenith and horizon. Presumably it is a cloud of very tenuous glowing gas. But we do not know whether it has any physical connexion with Spica, which is a fairly ordinary star, a bit hotter and heavier than the sun, but in no way exceptional.

As soon as we can get photographs of greater accuracy at different wavelengths we shall know a lot. For the moment we do not even know what sort of gas is emitting the radiation or how far away it is. Once a sputnik is equipped with instruments for measuring the approximate intensity of ultraviolet radiation coming from different directions, and signalling it back, our views on the distribution of matter in our neighbourhood (by which I mean the region whence light can reach us in a thousand years or so) may be very greatly altered. We already know, from radio reception, of huge clouds of hydrogen in the Milky Way. We may find that there is a lot more gas between the stars than astronomers think likely at present. If so, they may have to revise their present ideas about the formation of stars.

When a cluster is formed it may fly apart at once, and some clusters have been detected doing so in the last few years, particularly by Blauw in Holland, and by Ambarzumian. If it does not fly apart, most of its members will be held together by their mutual gravitation. However, such an association is not stable like a double star or a planetary system. A pair of stars sometimes approach one another fairly closely, though actual collisions must be very rare. Still, as the result of a series of such encounters, any star may acquire enough speed to shoot out of the cluster. Shelley wrote of

> 'The maze of planets struggling fierce towards heaven's free wilderness'.

The struggle of the planets appears to be hopeless. However, Chandrasekhar has calculated how long it should take for a cluster to lose half its members by this method. Applied to the Ursa Major cluster his method gives about 200 million years. But this is perhaps a bit too long. For the lighter stars have a better chance of escaping, and we certainly have not yet catalogued them all.

In the south soon after sunset we see Trisanku (Crux) followed by Beta and Alpha Centauri, the latter being the third brightest star in the sky. It is bright because it is the nearest star to us, being only 1·32 parsecs away. It consists of three, or possibly four members. Two are of about the sun's mass. One of these is about 20 per cent more luminous than the sun, the other about half as luminous. They go round their centre of gravity once in 80 years, which is almost the period of Uranus round our sun. The third is a very faint red star which is nearer to us than the main pair. It is about 16,000 times as far from them as our earth from the sun, and must go round them in something between one and two million years. Stars in clusters are often much closer than this, and there is no sharp line between a multiple star such as 61 Cygni, another of our neighbours, which has five components, and a small cluster. Men classify. Nature bridges the gaps between man's classes.

51

Jyestha

This month is called Jaisṭha, because during it the full moon is near to the star Jyesṭha in the constellation Brischik, or, in European terminology, to Antares, or Alpha Scorpii. Jaisṭha is a much better name than June, because it refers to an observable fact, as do all the other Sanskrit month names. A few thousand years ago Indians were very keen observers of nature. What is more, when the Rishi*. Visvāmitra gave the Gayatri Mantra to the world, he made it clear that those who recite it have the duty not merely to admire the divine glory of Earth, Air, and Sky, but to think about them. If everyone who repeats this mantra would observe and think, India would today be leading the world in natural science, as it sometimes did in the past. Similarly, if Christians forgave their debtors, as they promise to do in one version of the Lord's Prayer, they would have all the benefits of communism without its disadvantages. I sometimes think that the formulae of the great religions are directed to remedy the prevailing faults of those who practise them. They do not do so very efficiently.

* Sage.

Jyesṭha is the brightest of the row of stars which vaguely resemble the body of a scorpion. It is the fifteenth brightest star in the sky. While some of the constellations are not natural systems, but merely groups of stars which happen to be in nearly the same direction, other star groups are not only in the same direction, but at much the same distance. What is more they are moving nearly parallel to one another, so they formed a group a million years ago, and will still be a group a million years hence. Such stars are called a cluster. The easiest feature of two stars to compare is what is called their proper motion. This means the angle through which they move per year, against the background of faint and distant stars. According to the best data available seven years ago, Jyesṭha was moving through 32 thousandths of a second each year, in a direction 12 degrees west of south, while Shaula, the star which represents the scorpion's sting, was moving due south through 36 thousandths of a second. These figures will probably be known much more accurately when careful observations have been made for another century. From a human point of view this is a very slow motion. Jyesṭha will only seem to have moved through an angle equal to the moon's diameter after 60,000 years.

One can also find the speed with which a star is moving towards the sun or away from it by careful measurement of the wavelength of the lines in its spectrum. These are displaced towards the violet if the star is approaching us, and towards the red if it is going away. Shaula does not seem to be changing its distance, and Jyesṭha seems to be approaching at a rate of 3 kilometres per second. This is a small speed compared, say, with that of Rohiṇi (Aldebaran) which is increasing its distance from us by 54 kilometres per second. No doubt Jyesṭha and Shaula are not moving at quite the same speeds or directions, but the differences may partly be due to error, and partly to the fact that there really are motions inside a cluster.

The distances of the nearer stars are measured by a simple principle. We know the size of the earth's orbit in several different ways. One is by measuring the time taken for a radio signal to be reflected back from Sukra (Venus). The most accurate, till recently, was the blink principle, or parallax, as it is called. If we photograph Mangal (Mars) at the same time from Greenwich and Kodaikanal it will be in a slightly different place against the background of distant stars. The reason is the same as that for which a near object, say a chair in my room, seems to move relative to more distant ones, such as the far wall, when I look at it with my two eyes in turn. We know the distance from Kodaikanal to Greenwich, so we can calculate the distance of Mangal, and hence the diameter of the earth's orbit round the sun. Next, instead of two observations on earth, we photograph Jyesṭha from the earth in March and September. The two 'eyes' are 300 million kilometres apart. But even so Jyesṭha only seems to have moved through 16 thousandths of a second in six months. Some of the other members of the cluster move through over 20 thousandths of a second, others are further off, and have smaller parallax. Seven years ago a hundred and ten stars in the constellations Scorpius and Centaurus had been assigned to this cluster. The number is now probably a bit larger.

The parallax of all the bright stars has been measured, though some are too far away to give any definite result. Astronomers are constantly searching for fainter stars close to us with what is called the blink microscope. Two photographs are taken at an interval of six months, and arranged so that the images of most stars coincide. When looked at in rapid succession the images of near stars seem to move. The parallax of 8 thousandths of a second means that light takes about four centuries to reach us from Jyesṭha, having started in the time of Akbar and Elizabeth I.

Until the nineteenth century no one knew of what the stars were composed. Many philosophers, both in Europe and India, thought they were made of a different sort of matter, or *prakṛiti*, to that found on our earth. Then when spectral lines were identified with those observed on earth, it became clear that they were made of the same elements as are found here, but extremely hot, in fact consisting not merely of gas but of ionized gas. As a solid, such as iron, is heated up, it first begins to glow with red light, and becomes yellow, white, and blue as the temperature rises. The surface of blue stars is also hotter than that of red ones. Jyesṭha is a red star. Its Greek name Antares means 'like Mars', which is also red, for a different reason. It is in fact the reddest among the hundred brightest stars, except for Ardhra (Betelgeuse) in Mrigasira (Kalpurush or Orion). This means that its surface, or more accurately the layer from which most of its light starts, is relatively cool. It was Meghnad Saha, of Calcutta, who first showed how to measure the temperatures of stars from their spectra. The surface temperature of Jyesṭha seems to be about 3,000°C, as compared with nearly 6,000°C for the sun, and 40,000°C for very hot stars such as those forming Orion's belt.

When we know how far off a star is we can calculate its absolute luminosity, that is to say how much light it produces per second. If, for example, Jyesṭha were as near as our closest neighbour, Alpha Centauri, which one can see in the south at present, it would be of 9,000 times its present brightness, much brighter than Sukra, but not so bright as the moon even in her first or last quarter. Now most red stars are very dim indeed, too dim to be seen without a telescope even when quite close to us. In fact Jyesṭha is 100,000 times more luminous than the vast majority of red stars. There are no red stars in the intermediate range, whereas the gap between bright and faint yellow stars is much smaller, and there is no gap for the very blue stars.

We can measure the mass of a star when it is one of a pair revolving round their common centre of gravity. Many of the small red stars are double, so their masses are known, and are generally about a third of that of the sun. Jyesṭha is a double; but its companion is about 600 times as far from it as the earth from the sun, so its rate of movement has not yet been measured. However, for other reasons we think Jyesṭha is about sixteen times as massive as the sun. As it is cooler, each square centimetre of its surface is much less bright than the same area of the sun. But its surface is enormously greater, so it gives several thousand times as much light. In fact its average density is less than a millionth of that of the sun, and its bulk is so great that if it replaced the sun it would probably extend beyond the earth's orbit. However, its centre is almost certainly denser and hotter than that of the sun. The layer from which the light reaches us is almost empty, in fact by human standards nearly a vacuum. Outside it is a layer millions of miles deep in which not only atoms exist, as in the luminous layer, but also molecules built of two or more atoms. These stop light in a much more complicated way than atoms, and give rise to what are called band spectra. The molecule which produces the most striking change in the colour of the light from Jyesṭha coming up through it is titanium monoxide.

All along the Milky Way there are nebulae which consist of clouds of gas and dust. There seem to be three main types, dark nebulae which blot out the stars behind them, reflecting nebulae, in which starlight is reflected from dust, and emission nebulae, whose gases glow with visible light when excited by ultraviolet radiation from stars. Some of the dark ones appear to be condensing into clusters of stars. But we may have to wait some centuries before we know whether the 'globules' seen in some of them, which are much larger than our solar system, do in fact collapse under their own gravitation to form stars. This is, however, how stars are now thought to originate. Their masses are mostly less than the sun's, and Jyesṭha is unusually massive. As the matter, which is mainly hydrogen, falls inwards it heats up until nuclear reactions can begin. The rate at which energy is liberated increases roughly with the cube of the mass. Jyesṭha probably started its career not much over 30 million years ago as a 'supergiant' blue star about 10,000 times as bright as the sun, and thus used up its hydrogen, which was mostly converted into helium, 1,000 times as fast. When the hydrogen was largely exhausted the centre got hotter, and it began to convert helium into heavier elements. The whole star expanded enormously. Whereas Shaula is only about eight times as massive as the sun, and has only about eight times its radius. It is believed that it will swell up and redden later. Possibly Jyesṭha will swell still further, but it is not likely to get much brighter. It may become a variable with a period of several years. It may then contract again and get whiter, with a shorter period. But at some stage it appears quite likely to explode. If it does so it will be visible by daylight for several days, if any men are left on earth to see it. Perhaps the 'life' of such a massive star, from formation to explosion, is a matter of 50 million years or so. Whereas our sun seems to be about four thousand million years old, and may last very much longer.

Such predictions are based partly on calculations made with electronic computers on the basis of physical measurements made in laboratories on earth, and partly on statistics as to the numbers of stars of various kinds found in different clusters. But the mathematicians are much more confident about the evolution of ordinary stars like the sun than about very heavy ones.

The cluster of which Jyeṣṭha is the brightest member is moving relative to the sun. If a flock of birds flies in a straight line and keeps together, it will gradually appear smaller and smaller, and finally seem to contract to a point. The Scorpius–Centaurus cluster will apparently move right across the southern sky to the constellation Puppis, south of Canis Major. But on its journey it will lose some members which have developed high enough speeds from approaches to other stars to escape from the gravitation of these neighbours. This is much more likely to happen to light than heavy stars, and Jyeṣṭha will probably stay in the cluster till it explodes.

I am quite aware that some of what I have written is speculative. In particular the past and future of stars and star clusters is based on calculation and not on observation. However, there are several different methods of determining their distances, speeds, and sizes which agree remarkably well. I am not an astronomer, but till I came to India I tried to keep up with astronomy. Now I am some years out of date. But the subject still fascinates me. This article was largely taken from Allen's '*Astrophysical Quantities*', which, along with the poems of Virgil, Milton, and Blake, is one of my favourite bedside books.

I have just been reading Professor Sadhan Kumar Ghosh's '*My English Journey*'. It has, I am told, been criticized. I think its English style, with its curious mingling of classical diction and modernisms such as 'gimmick', is admirable, and that many native English writers could learn from him. Although I naturally disagree with some of his estimates of English people and institutions, I think none of them is unfair. His worst misjudgement is, I believe, his statement on page 112 that 'Camus, Sartre, and Beckett arouse a shudder of horrified pity for the whole of humanity.' On the contrary, I think Sartre is the one living author who can depict 'plain heroic magnitude of mind'. Goetz and Canoris, the leading character of '*Morts sans sepulture*', in my opinion the greatest drama of this century, are real heroes. No doubt they came to violent and painful ends. But I am proud to belong to the same species as Canoris.

On page 35 Professor Ghosh quotes from a translation of a poem in the Greek anthology attributed to the great astronomer Ptolemaios (or Ptolemy). A somewhat more accurate, if less poetic, translation than his own runs as follows:

> 'I know that I am mortal and ephemeral, but whenever I am tracking out, according to reason, the back and forth spiral tracks of the stars, I no longer touch the earth with my feet, but am fed by Zeus himself with god-nourishing ambrosia'.

'Ambrosia' is the same word as amrit. Zeus had most of the attributes of Indra. Two different Greek texts give slightly different versions of the astronomer's mental process. The tracks of the planets in relation to the 'fixed' stars are, of course, spirals. I do not hesitate to say that this poem and the Gayatri Mantra each help me to understand the other. Ptolemy's poem is a good description of the feelings which I get several times a year when I or one of my colleagues make a scientific discovery of a really unexpected kind.

Professor Ghosh quotes the first and last phrases of this poem. He omits any reference to astronomy. If the emotions of the last phrase had been aroused by a woman, and the Professor had omitted the fact, I might have used a rude word like 'castration'. If the poem had been inspired by a deity I could have written about atheistic bias, and so on. But if a poet is inspired by scientific research, this is a fact to be suppressed by Professors of Literature. To quote Professor Ghosh's page 82, 'One of the saddest things in the English Universities today is the emergence of the Literary Specialist'. It is sad in India as well.

Visvāmitra was a great poet. Very likely he experienced the Gayatri Mantra as a revelation. This is the usual experience of scientists who arrive at new theories. They certainly do not produce them consciously, and the process feels like a discovery rather than an invention. Visvāmitra, I suspect, was also an astronomer. I do not believe that he created new constellations such as Trisanku. I think it very likely that, after travelling south, he described them to people in northern India who had never seen them. I should like to see a revival, in modern India, of the emotions which, I believe, animated him. This is why I have written this article.

52

Simplifying astronomy

In my last article on stars I wrote about Sirius, the brightest star in the sky, and one of the nearest. Since I did so this star has been very much in the news. A British and an Australian radio expert, Hanbury Brown and Twiss, have revolutionized one side of astronomical technique. Without any giant telescope, and indeed without any optical apparatus but two second-hand searchlight mirrors left over from the last war, they have succeeded in measuring the diameter of Sirius. Their apparatus included a lot of electronic components, but these are available in India. The principle employed was that of interference. It has long been known that when light from the same source shines through two fine parallel cracks very close together and then falls on a screen we do not merely see two white lines. If the light is monochromatic, that is to say all of the same colour and wave length, we get a series of parallel light and dark bands. The paths from the source to the centre of a light band differ by a whole number of wave lengths, so that waves coming through the two cracks reinforce one another. The paths to the centre of a dark band differ by a whole number of wave lengths plus a half, so that the wave from one crack tends to pull electrons in one direction, that from the other in the opposite direction, and they cancel out.

The precise pattern of interference bands depends on the wavelength of light, the width of the cracks, and their distance apart. The principle can be used to measure the diameter of a star. Because the star is not a point the pattern differs from that which would be obtained if it were one. The principle of interference was used by Michelson and Pease about thirty years ago to measure the diameter of a few very large red stars like Betelgeus, the bright red star in Orion. But their apparatus was very expensive, and could only be used on a few stars.

Twiss, who works in Australia, proposes to use the new apparatus on the star called Suhail el Muhlif, or Gamma Velorum, which is one of the most interesting stars in the sky. At present [March 1959] Canopus, Agastya, or Alpha Carinae, is conspicuous in the south about sunset. This is a very distant star, but produces so much light that it is the second brightest of all stars in the sky, even though if the sun were removed to the same distance we should not be able to see it with the naked eye. About an hour and forty minutes after Canopus two fairly bright stars reach the southern meridian. These are Gamma Velorum and Zeta Puppis almost due north of it. These are the two hottest stars which appear bright enough to be easily recognized. More accurately Gamma Velorum is a double star consisting of two components moving round their common centre of gravity of which the hotter and brighter is what is called a Wolf–Rayet star (after the names of two astronomers). Wolf–Rayet stars show a continuous spectrum, corresponding to a surface temperature of about 80,000 degrees, and also bright lines. In fact their spectrum is superficially like that of a gas discharge lamp with a fluorescent background.

Most astronomers think that the continuous spectrum comes from the surface of the star, and the bright lines from an atmosphere surrounding it. The new technique will enable astronomers to measure separately the diameter of the star itself and of the atmosphere round it which may be ten times greater. As Gamma Velorum appears overhead in Australia, Twiss is probably measuring it now. But as the new techniques can be applied to any stars bright enough to be seen, there are plenty of opportunities for Indian astronomers or physicists to use them on other stars without spending much foreign exchange.*

The stars with very hot surfaces are believed to be relatively young. In a few cases where several type O stars like Zeta Puppis are moving away from a common centre, we can calculate the date of their origin, at which they started from it. Some of the stars in Trisanku, or the Southern Cross, which is conspicuous about midnight at present, are also pretty hot, and must be fairly new. Now according to Indian tradition, the great rishi Visvamitra made several southern constellations, including Trisanku. I find Visvamitra, who had considerable difficulty in controlling his passions, a good deal more sympathetic than many men of a holier disposition who found the path to enlightenment easier. But I am afraid I don't believe he created new star systems. I want to suggest that he may quite possibly have noticed the appearance of new stars in the southern sky, and pointed them out to his contemporaries. If so the legend might easily arise that he had created them, especially if they increased in brightness for some years after their first detection. We know much too little about stellar evolution to say with certainty that this is possible or impossible. Whereas we can say that, for example, the Seven Rishis are many millions of years old, and were shining long before there were human beings. I hope to write about them in a later article.

I want to draw the moral that Indian scientists would not complain so much about lack of equipment if they would keep

* In the three years since this was written, I have heard of no evidence that they have done so.

their eyes open for cheap methods invented in other countries. It is of course still better to invent cheap but efficient apparatus oneself, as J. C. Bose, for example, did. I have just induced the Indian Statistical Institute to offer a post to a colleague who has done this. Personally I try not to use apparatus at all, and to get my junior colleagues to work with as little as possible. One reason for this is a moral one. The use of complicated apparatus separates scientists who use it from ordinary men who use such simple machines as ploughs and potters' wheels. There is in fact a very great deal to be found out about such machines. In particular what are the best designs for a plough to be used in the various types of soil found in different parts of India. I am pretty sure that most of the existing kinds demand a bit more work from a bullock than is absolutely necessary. I think the physics of ploughing is more important for India than some of the branches of physics which are being investigated in our university laboratories, and quite as interesting scientifically. I am not a consistent Gandhian, but I certainly think that Indian scientific research would be the better for adopting a few Gandhian principles, one of which is to regard machine as made to serve men, and never to think of men as made to serve machines.

53

New light on memory

Our knowledge of the relation between mind and brain grows very slowly. There are two reasons for this. We cannot ask the right questions. Whatever may be the nature of the relation it cannot be the same as that between two material objects, for example water in the jar which holds it, or between two conscious beings such as a ruler and a subject. So the kinds of precise thought which characterize most of the sciences on one hand, and law on the other, will not help us. I think it very possible that the minds of human beings can be regarded as illusions resulting from ignorance. However, we must describe these illusions as best as we can.

Secondly, we have not the right to experiment with human minds or brains as freely as with other parts and functions. Even if neurologists were permitted to injure the brains of condemned criminals I do not think they would do so. A neurologist who did not realize that an executioner is at least as severely damaged as the man whom he kills would be so bad a psychologist as to be unlikely to find out anything of interest.

So, although we can learn something from experiments on animals, which, however, often cause some damage to the experimenter, much of our information comes from brain surgery. Fortunately for the progress of science this generally includes some experimentation. Except in the rare cases where the operation is merely the removal of a tumour or a splinter between the skull and the brain, it is part of the practice of brain surgeons to stimulate the exposed brain surface electrically. For example, if the operation is being done to cure the unfortunately rare type of epilepsy which can be so cured, a part of the brain is stimulated so as to bring on epileptic fits, and the most sensitive region is cut out. During and after any cutting of the brain photographs are taken, and the tissue removed is examined under the microscope.

Brain operations are now seldom done under a general anaesthetic. This is partly because they often last a very long time, and many people who recover after an hour of ether anaesthesia would not recover after six hours. It is partly because an important function of anaesthesia is to cause relaxation of muscles, for example in the abdominal wall, and this is not needed in brain surgery. And it is partly because a conscious patient can give the surgeon important information. So the skull and other tissues outside the brain are numbed with a local anaesthetic.

Cutting or electrical stimulation of the brain surface never causes pain. Pain is felt with a part of the brain far inside it, and at present inaccessible to surgeons. Some parts of the brain surface are concerned with muscular movements. In human beings the hand muscles, the eye muscles, and those concerned in speech, occupy most of the area. For these are muscles over which we exercise far more detailed control than over those of the trunk and legs. Electrical stimulation of these areas causes involuntary movements. Near to the motor areas are the main sensory areas. Again the hand is represented by as large a brain area as the whole trunk. Stimulation of these areas gives rise to feelings of touch or numbness in the corresponding parts of the body. Vision has a special area at the back, and hearing at the sides. However, most of the brain surface is 'silent'. Its stimulation produces no obvious effect. Areas near the main sensory areas are concerned with coordinating sensations. Thus when a silent area near that whose stimulation causes sensations in the hand has been damaged by a bomb splinter, the patient can still feel touch on any part of his hand, but cannot tell whether he is grasping a coin or a match box.

This has been known for forty years. More recently an area just behind the forehead has been shown to be concerned with personality. If the fibres in it are cut, violent and intractable patients may become quiet and cooperative. One particular part of it, the cingular gyrus, seems to have a very remarkable function. This has been removed in some patients dying of a very painful inoperable cancer. The patients say that they are still aware of severe pain, but it does not distress them. Nor does the thought that they will be dead in a month. It would seem reasonable to say that this area was the seat of what Indian philosophers have called *ahankara* the sense or illusion—of personal identity. Naturally enough, operations or injuries in this region may have very undesirable effects. Some victims appear to lose initiative completely. For example, a woman could not decide on what to serve as a meal, but could cook it satisfactorily if someone else made the decision.

The most recent addition to our knowledge is described in a paper read to the Washington Academy of Sciences last year by the eminent brain surgeon Penfold, who was one of my fellow workers in Sherrington's Laboratory in Oxford thirty-seven years ago. In the course of operations on fully conscious patients he has frequently stimulated areas on both sides of the brain which had previously been regarded as 'silent', with the following results. Sometimes the patients were conscious of an event in their past in very great detail. For example one woman heard a musical piece being played by an orchestra. She thought it was coming from one of the pieces of electrical apparatus in the operating theatre and was able to hum some of the tunes. She remembered when and where she had heard it before. Others lived again through emotionally important events in their lives. The second effect was a feeling of recognition of their immediate perceptions as having been experienced before. These two experiences are essential components of human memory, though Penfold never elicited both of them together.

I meet a man whom I have not met for several years. I recognize him as someone whom I have known before. I then remember that when we last met he was wearing a gold-edged lungi* and told me about tobacco growing in Madras State. I may have the memory of past events before or after I have the recognition of the face, but both are needed for satisfactory memory. I am not satisfied if as is often the case, I merely remember that I have met the man somewhere before. In English we use such words as 'memory' and 'remember' very loosely. Thus I may say that I remember a piece of verse, meaning that I can recite it, I do not necessarily remember when, where, or how I learned it. In other languages the distinction between true memory and capacity for a performance is more clearly made. If we learn more about the physiology of memory we shall probably be able to make more useful distinction between the various meanings of this word.

Penfold does not think that either the recognition of present experience or the recall of past ones is due to activity in the part of

* South Indian sarong.

the brain which he stimulated. As a general rule, electrical stimulation of a part of the brain does dot give a detailed action or sensation, Electrical stimulation of the speech centre may force the patient to produce a prolonged vowel sound, but not a sentence or even a word. This is natural enough, for a great many cells are stimulated at once, and there is no proper pattern of action. But stimulation of one part may occasionally evoke a clear pattern of action or sensation in another part. Penfold, therefore, thinks that the actual seat of memory is elsewhere. He has merely stimulated nerve fibres which send messages to another part of the brain which bring stored memories into consciousness. I think he is probably right, but most of the facts about cerebral localization can be interpreted in a less materialistic way, as Bergson tried to do in his '*Matière et mémoire*'.

What is quite certain is that philosophers ought not to ignore these facts. If, for example, they hold to analysis of the human mind given in the samkhya philosophy, they should at least realize that the different constituents of the mind described in that philosophy are associated with different parts of the brain. Lenin wrote as follows: 'It is, of course, totally absurd that materialsm should maintain the "lesser" reality of consciousness or should necessarily adhere to a mechanistic world picture of matter in motion and not an electromagnetic or even some immeasurably more complicated one.' My own guess is that our most useful world picture a century hence may include at least some elements of the samkhya picture, matter being a good deal more like prakiti according to Kapila than like the picture common when Lenin wrote. But we shall certainly not reach a true picture by ignoring facts such as I have summarized in this article.

54

Relations between biology and other sciences*

The relation between two sciences should be one of mutual aid. A science advances both when a new question is asked and when it is answered. Thus astronomy advanced when Hipparchus asked 'Do the so-called fixed stars change their positions relative to one another?', and first catalogued such positions. It was a greater advance when proper motions were detected and measured.

Biologists certainly ask questions of other scientists, such as 'Can you make me an instrument to see the detail of objects less than .001 mm in diameter?' 'What is the composition of the wax produced by this plant?' 'How many years have elapsed since our ancestors became land animals?' But workers in other sciences can and should ask questions of biologists. Perhaps the chemists and geologists have the most important questions to ask.

Mathematics

Let us begin with the relation of biology with mathematics, the most abstract of the sciences. Some of the questions which the biologist asks are really questions of physics. For example, in the Proceedings of the Royal Society for 1951 G. I. Taylor discusses the swimming of microscopic flagellate organisms such as many algae, protozoa, and spermatozoa. Here the viscosity of water is its most important property, and the mathematical problem is quite different from those raised by swimming organisms even a centimetre long, let alone ships or aeroplanes. Taylor has only solved the problem in two dimensions,* so there is plenty more to be done.

* The full text of Professor J. B. S. Haldane's lecture delivered on 2 January 1952 at the 39th Session of the Indian Science Congress Association held at Calcutta.

There are, however, a number of mathematical problems raised by purely biological facts, particularly problems of population. The rate of growth of a population depends on its past. Thus that of a human population depends on the numbers of women at various ages between 15 and 45. Such problems inevitably lead to integral equations. In fact the modern theory of integral equations, and of functionals in general, owes much to Volterra's investigations on fish populations. These have in turn reacted on biology by suggesting biological characters of organisms which should be measured, for example fertility as a function of age. My own investigations in this field have led to some novel solutions of non-linear finite difference equations.

The branch of mathematics which owes most to biology is statistics. The normal or Demoivre-Laplace-Gauss distribution was shown by Quetelet to apply to human height and other biological measurement, and the theory of correlation, so important in economics, was largely developed by Karl Pearson in his investigations on heredity. The benefit has been mutual. Biometry is now an essential part of biology.

Another branch of mathematics which is growing in connection with biology is cybernetics, defined by Wiener as the theory of control and communication in animals and machines. Because this theory has been developed in connection with machines, for example thermostats, it is sometimes thought that an attempt to apply it to men and animals involves the assumption that they too are machines. This is incorrect. An animal may agree with a machine in a mathematically determinable characteristic, for example the rate at which its heat loss increases when its temperature rises; just as it can agree with the machine in weight, horse power, and so on. We can then legitimately argue from the machine to the animal without equating them.

Physics

To illustrate the interactions of physics and biology, I will mention just two fields of physics, the new subject of nuclear physics and the old one of polarized light. Physicists have now given us a time scale determined by radioactivity, which is long enough for the very slow process of evolution. It is, however, worth mentioning that it was a biologist, Darwin, who first gave nearly the correct time scale for geological events. In the first edition of *The Origin of Species* he calculated that about 300 million years (or 70 *mahakalpas*) had elapsed since the end of the Jurassic period. The correct number is about 120 million, but Darwin's contemporaries were thinking in terms of thousands of years for the whole age of the earth, and even those who were most emancipated from ancient tradition gave the total age as 10 or 20 million years. Darwin unfortunately withdrew this estimate in later editions of his book.

Biologists use nuclear physics mainly to trace the path of an atom through the maze of metabolic processes; for example a man is fed with glycine containing a heavy isotope of nitrogen. This appears in the pyrrol groups constituting the prosthetic group, haem, of his haemoglobin. If other amino-acids with 'labelled' nitrogen are fed very little appears in the haem. This experiment may be compared with the labelling of a number of ducks in winter in India by rings on their legs. A fair number of these ducks are later found in Siberia, a very few in Germany, and so far none in Britain. Again, Urey finds that the ratio of the two oxygen isotopes in the calcium carbonate of modern molluscan shells is mainly, if not wholly, determined by the temperature at which they were formed. Applying this result to shells from the English chalk, he finds that they were formed at temperatures up to 24°C, as warm as the Indian Ocean today. This method may yet solve one of the at first sight insoluble problems of evolutionary biology, namely the date at which the ancestors of mammals achieved temperature regulation. Biology has not as yet contributed to nuclear physics.

Now consider a phenomenon known for more than a century, namely the polarization of light. Von Frisch has quite recently shown that bees can perceive the plane in which light is polarized, and used the information to find their way when a patch of blue sky is present, since the light from the sky is polarized. Other insects and arachnids have been shown to share this power. How far do other animals perceive physical events which do not affect human senses? Certainly some fish perceive electrical pulsations, a good many insects perceive radiation in the near ultraviolet. But there is no evidence that any animals perceive magnetic fields, though such perception would aid them in migration. Biologists are also using polarized light for their own purposes. For example, Swann and Mitchison have used it to detect both true crystallization and less complete orientation of protein molecules in living protoplasm, and to measure the thickness of cell membranes. For this purpose a really efficient polarizing microscope is needed. Its design is very different from that of an ordinary microscope. For example, the latter includes compound lenses of different types of glass to correct for chromatic aberration. But the interfaces in such lenses partially polarize light passing through them, and are thus undesirable. In fact the whole theory of microscope design has to be revised, and a number of interesting optical problems arise. Here, then, biology is contributing to practical and theoretical physics.

Chemistry

The relations between biology and chemistry are very intimate. Organic chemistry began as the chemistry of biological products, but it soon achieved independence, developing its own synthetic methods. Biochemistry differs from organic chemistry both in its methods and its aims. It uses biological methods for the detection and estimation of substances which are active in very small concentrations, such as vitamins, hormones, and antibiotics; and it uses special physical and chemical methods for the separation and study of large molecules such as proteins. One of its main aims is the study of chemical reactions occurring within living cells. I have already lectured to the physiological section of the Congress on the use of genetical methods in this study. The reactions are often very different from those devised by organic chemists, and far more efficient. A striking feature of them is the frequency with which phosphoric esters occur as intermediates. I have little doubt that organic chemists will gradually come to copy these reactions, using highly specific catalysts giving a yield of 99 per cent or more, rather than less specific processes giving yields of often under 50 per cent.

But rather than develop this theme I want to suggest to you that certain problems of chemical engineering applicable to plant products may literally be matters of life and death for India. India is not producing enough food for its population. Some people say that it is overpopulated, and that no improvements in agriculture can keep pace with the probable increase of its population. Let us look at the problem from a rather different angle. All or almost all, sources of human food are plants or animals which could be used by primitive man, but whose yield has been increased both by selective breeding and improved agricultural methods. But in industry we are not content with using substances such as wood, iron, and glass which have been known for thousands of years. We also use others, such as aluminium, rubber, and plastics, whose very existence was unknown 200 years ago. Is it possible to use sources of food which were not available to primitive men? I believe so. Men cannot digest grass or the leaves of most trees. They contain too much indigestible fibre, not to mention silica. At present our most efficient way of utilizing grass is to get a cow to eat it, and give us back a fraction of its proteins and carbohydrates in the protein, fat, and carbohydrate of her milk.

However, in the laboratory it is quite possible to separate most of the proteins and the digestible carbohydrates of grass and other leaves from the fibre. While seed proteins generally contain much glutamic acid and are often short of tryptophan and other constituents needed by men, leaf proteins have a much more favourable amino-acid composition. But their separation from fibre is difficult even on a laboratory scale, and at present quite impracticable on a factory scale. A few people are working on this problem in England, but our chemical engineers are more interested in the mass production of explosives than of food. India will get no help from the Soviet Union or the United States. Both countries have plenty of undeveloped land, and ordinary agricultural methods will suffice them for a generation to come, at least. Even if you begin to work on the problem in India tomorrow you would get no substantial increase in food production for 10 or 20 years. But I believe that India ought to look that far ahead.

Agricultural science

My own science of genetics is, of course, intimately connected with agricultural science. In this field India cannot rely on other countries, as it can to some extent in physics, chemistry, and engineering. The strains of rice which give high yields in Italy and Japan cannot be expected to do so in India. However, Dr Mukherjee has dealt with problems of plant improvement in his presidential address to this Congress. I shall confine my remarks to the improvement of animal races. Let me take a simple example of the type of work needed. The villagers round Calcutta keep ducks and hens. It is possible to breed hens which can produce a large weight of eggs per year, but to do so you must feed them on grain, so that they compete with men for an important food. On the other hand ducks live wholly or mainly on animals and plants in the tanks, and do not compete with men for food. However, these ducks have not been selected for high egg production, though attempts are being made to acclimatize English ducks at one place in Bengal. My wife has been studying your domestic ducks, and finds that they differ greatly from those of Europe or of China. It may turn out that our European ducks are no more adapted to life in Bengal than are our cattle. If so you will have to form your own breeds, and even in five years a good deal of progress could be made.

Your great problem is, however, of course, the cow. Some people will tell you to kill and eat the old cows and most of the male calves. As a biologist I would say something different. The relation between the Hindu and the cow, at its best, is the best example known to me of symbiosis between two vertebrate species; at its worst, it leads to starvation of men and cows alike. If a people decides to adopt a symbiotic rather than a predatory relation with another species, this raises special biological problems which only they can solve. We in England have got a great deal more milk to drink per man than you in India. This is because each of our cows gets a great deal more to eat than the average Indian cow, and has been selected so as to give a high milk yield if well fed. You will have to breed for higher milk yield, while reducing your cattle population so that each cow gets enough to eat. I see from the programme of your Congress that you are studying methods to render cows of uneconomic type infertile. I am glad that your ethics permit this. For the moment you should certainly breed mainly for high milk production.

But two other aims must be considered. If you are to reduce your cow population without killing cows, you will ultimately have to increase the duration of lactation so that you do not have to breed so often from each cow. Some European breeds stay in milk for over 600 days, and even this number could be increased, perhaps by selective breeding, certainly by hormone injections; though the latter method is not at present economically sound. It may well be that you will also need fewer bullocks. This will certainly be so if you mechanize your traction. If so, you will want to be able to control the sex ratio of your cattle, diminishing the proportion of male to female calves. The sex ratio can be controlled in some insects, and in several amphibian species. For example, by hormonal treatment one can induce an animal which would otherwise be a female to become a functional male, and such an animal has an excess of females in its progeny. If the human race decides to use up its oil supplies in the next fifty years it will be more important to breed a race of cattle combining high milk yield in females and great muscular strength in males. None of these ideals is unattainable, but only Indian biologists will solve them, since in other countries the flesh of cattle is eaten, and bullocks are less used as a source of power. If, however, Indian biologists can solve them they will greatly increase the productivity of their country, and perhaps set an ethical example to the world.

Geology

The relations of biology and geology are, of course, intimate, for palaeontology shares the methods of both sciences, and gives the biologist his most important data for the study of evolution, while allowing the geologist to determine which rocks were laid down synchronously in different areas. There are great opportunities for palaeontology in India, of which I mention only one. Our knowledge of Jurassic mammals is extremely small, largely because most European and North American Jurassic strata are marine. But India possesses terrestrial upper Gondwana beds of Jurassic age, notably the Jabalpur and Raghavapuram beds. These offer a unique hope of finding early mammalian remains.

As an example of the aid which biology can give to geology I take Ma's work on Devonian corals. Living corals form well-marked annual layers if they live far enough from the equator, but do not do so if they live near it, where seasonal temperature changes are slight. Ma considers that the Devonian equator was a good way from the present one, running, for example, through Europe, but that owing to the drifting apart of Europe and North America, it does not form a great circle on the earth's present surface. If this is correct continental drift has occurred. Similarly, from the present distribution of several types of animals, but particularly torrential fish, where the animals of the Western Ghats appear to be derived from those of Burma, Hora has deduced the Satpura hypothesis. According to this hypothesis, there were fairly recently, perhaps in the Pleistocene, hills in what is now lower Bengal, connecting the hills of peninsular India with those of Burma. Whether or not this hypothesis is finally accepted, it has led to researches, both biological and geological, whose results are of value for both sciences.

Anthropology

Physical anthropology, which regards men as animals, is a branch of biology. In the last generation its technique has been revolutionized by the discovery of a set of characters, namely the antigens of the blood corpuscles, which are wholly determined genetically, and not at all influenced by the environment, as, for example, skin colour and stature certainly are. Unfortunately Indian anthropologists have so far mainly confined themselves to the study of those antigens which are important when blood is transfused. Similar work with all the antigens whose genetics are known would certainly give new data on the racial history of India, as it has on that of Europe. In particular it would throw light on the relation of the primitive so-called aboriginal tribes to the majority of the population, whose ancestors were already civilized when most of Europe was in the tribal stage of culture. It is important that this work should be done as soon as possible in those areas where the primitives are beginning to interbreed with the rest of the population.

Biology can make an important contribution to the knowledge of man by the genetical analysis of domestic plants and animals. As the President of the Indian Republic recently said, human history should not be the story of violent deeds, but of cultural achievements of which we may be proud. The domestication of a plant or animal species is such an achievement. There can be no doubt that poultry, along with a fair number of plants, were first domesticated in India. Bees may have been so. It is much less clear whether, for example, cows and dogs were first domesticated here. It is up to Indian biologists to find out. One method would be to compare the blood antigens of dogs with those of jackals and wolves, a process, by the way, which does not involve the taking of life or the infliction of appreciable pain.

Logic and Moral and Social Sciences

But biology has a still more important relation to the human sciences to whose importance Pandit Nehru referred in his address this morning. If I am asked what is the greatest scientific achievement of the century, I shall point to the great advances in atomic physics associated above all with the names of Rutherford, Planck, and Bohr, but to which Raman, Saha, Bose, and Kothari, to mention no more Indian physicists, have made honourable contributions. Most people, alike in Western Europe, the USSR, and the USA would agree with me. So, rather to my surprise, did Mr. Justice Banerji this morning. But for thousands of years Indian sages have said that the most important thing that a man can do is to realize, in thought and action, the unity of all living, or at least of all sentient beings. Sometimes they expressed this unity in terms which I can no more accept than I can accept European mythology, sometimes in more abstract terms such as 'Tat twam asi'. My own philosophy is decidedly *nastika*, but I am not so dogmatic as to deny the possibility that the Vedanta, or one of the other *astika* philosophies, may be nearer to the truth than my own.

If this is so, then the greatest achievement of nineteenth-century science was Darwin's statement, in highly concrete terms, of the unity of life. And the greatest achievements of twentieth-century science are the discovery by biochemists of the great similarity, in some cases the identity, of the chemical processes of life in animals, and even in plants, and perhaps above all von Frisch's discoveries in the field of animal language. You will remember that he has shown that one bee can make to others such a statement as this: 'there is nectar requiring at least 20 gatherers in apple blossom at 900 meters 12°E of N from our hive'.

It communicates this information by dancing. The direction of the dance on the comb indicates the direction of the flowers, an upward dance indicating the direction of the sun, an oblique dance 30° to the right of vertical, a direction 30° to the right of the sun, and so on. The rhythm with which the abdomen is oscillated indicates the distance, the duration of the dance, the amount of food, while the other bees smell the dancer to determine the nature of the flowers from their scent. Now so far, only the hive bees of Europe and North America have been investigated. Is the language the same for Indian domestic bees, and above all for Indian wild bees, which may represent the ancestors of all domestic races? This problem can only be solved in India. The study of bees is of considerable economic importance. I have seen no domestic bees in Bengal, though there are plenty of flowers. There may be good reasons why bees cannot be kept in some parts of India, but they could certainly be kept over wider areas than at present. Not only would they add to the food supply without taking food from men or cows, they would also improve the fertility of many fruit trees.

Similar work has been done on bird languages in the Soviet Union. These languages are mainly learned, while that of bees seems to be biologically inherited. In Ukraine one species of tit sings two different languages, and the two linguistic groups do not interbreed. Such investigations seem to me as essential a background for social anthropology, for philosophy, even for logic, as is comparative anatomy for human anatomy. In so far as logic tells us anything about the nature of reality, and not merely about human communication, the laws of logic must apply to dancing as well as to speech. It is, however, clear that dancing, as a method of communication, suggests a 'many valued' logic rather than one with a sharp distinction of true and false.

Mr Justice Banerji told you this morning that whatever progress might be made in positive science, it could not satisfy the soul. May I venture to differ from him? I believe that the contemplation of living beings can do so. Let me quote three sentences from Francis Darwin's reminiscences of his father, Charles Darwin: "I used to like to hear him admire the beauty of a flower; it was a kind of gratitude to the flower itself, and a personal love of its delicate form and colour". "I think he sometimes fused together his admiration of the structure of a flower and its intrinsic beauty." When observing animals he often stood completely still. "It was on one of these occasions that some young squirrels ran up his back and legs, while their mother barked at them in agony from a tree." Similar incidents of concentration are, of course, recorded of *rishis* in India.

I believe that we can learn moral lessons from animals. Apparently gentle animals such as doves may quarrel, but if so, in nature, one flies away before much harm is done. If however, they are kept in a cage one dove may slowly peck another to death. On the other hand, wolves and ravens fight with their formidable teeth and beaks. But such fights always end abruptly if the vanquished surrenders. Lorenz describes how a wolf which is beaten presents its throat to the victor so that a single bite would kill it. The victor growls, but is unable to bite. Similarly a vanquished raven presents the back of its neck, and the victor never kills it. Unfortunately men are not always so merciful as wolves and ravens. In the past both British and Indian soldiers have murdered prisoners, and today British airmen are dropping bombs on civilians in Korea.

I end on this note for a simple reason. It is entirely possible that within the next two or three years European civilization will largely destroy itself. If you in India continue your present foreign policy you may escape this destruction. If so I hope that you will carry on with the great traditions of our biology, rather than imitate us in self-destruction with weapons provided by the physicists.

55

Darwin in Indian perspective

It is probably too early to assess Darwin's significance for human culture. It is, however, much easier to do so if one has the stereoscopic view afforded by a measure of intimacy with more than one of the main cultures of our planet. I could not have written this article before I became an Indian. To Europeans and Americans it inevitably seems that Darwin's greatest achievement has been to convince educated men and women that biological evolution is a fact, that living plant and animal species are all descended from ancestral species very unlike themselves, and in particular that men are descended from animals. This was an important event in the intellectual life of Europe, because Christian theologians had drawn a sharp distinction between men and other living beings. In view of Jesus' remarks about sheep, sparrows, and lilies, this sharp distinction may well be a perversion of the essence of Christianity. St Francis seems to have thought so.

But in India and China this distinction has not been made, and according to Hindu, Buddhist, and Jaina ethics, animals have rights and duties. My wife has stated categorically that Darwin converted Europe to Hinduism. This is, I think, an exaggeration, but is nearer to the truth than it sounds. Hinduism is not a religion as this term is understood by the adherents of proselytizing religious beliefs. It is an attitude to the universe compatible with a variety of religious and philosophical beliefs.

Such attitudes are best shown in imaginative writing and art. In one of the two great epics of ancient India, the *Ramayana*, the divine hero, Ram, is aided to regain his wife, Sita, who has been abducted by the ogre Ravan, by an army of monkeys and bears, acting on information received from vultures. In a relic at Mahabalipuram depicting scenes from the life of Krishna, which is one of the masterpieces of medieval Indian art, the whole background consists of cows' heads. For every Hindu the background of human experience is alive. Of course he does not live up to his attitudes and beliefs. Nor do Christians. If even 50 per cent of Christians forgave their debtors, from the boss who owes a week's wages to the farmer who has mortgaged his means of livelihood, the economic fabric of Christian civilization would collapse in eight hours. Similarly many, perhaps most, Indians, are cruel to animals; but kindness to animals, including vegetarianism, is commoner in India than forgiveness of debtors in Christendom.

If Darwin had died young, Wallace would presumably have promulgated the theory of evolution by natural selection when he did, and it would probably have been accepted, though as Wallace's arguments covered a smaller field than Darwin's, the acceptance might have been slower. And as Wallace left loopholes open for supernatural intervention, which Darwin did not, the immediate effect on Western thought might have been less.

In my opinion, however, Darwin's most original contributions to biology are not the theory of evolution but his great series of books on experimental botany published in the latter part of his life. They are concerned with those aspects of plant life which are most like animal and human life. Two are devoted to climbing plants and insectivorous plants respectively and three to sexuality in plants, but particularly those aspects which are most human, such as the evil effects of incest (the theme of *Oedipus* and *The Cenci*) and the strange devices by which its most extreme form, self-fertilization, is avoided. The facts discovered are momentous. Among their applications are the discovery of plant hormones and the invention of the weed-killers which resemble them chemically, and the systematic outbreeding of maize, of whose importance for the agriculture of the USA I need not write.

But what was the attitude which led to these discoveries? To answer this question we must not only read Darwin's books, but his autobiography, and the memoir by his son Francis. Perhaps the most enlightening passage is Francis' account of what his father called 'a fool's experiment'. Francis was ordered to play the bassoon to some seedlings. In fact this did not influence their growth, as vibration of the table had done. However, other fool's experiments came off. Perhaps Darwin's classical fool's experiment was to cut a number of scalene triangles of paper, to leave them on his lawn, and find that the earthworms which used some of them to plug their holes generally chose the most acute angle to drag as far as possible down the hole. Darwin did not draw a sharp line between earthworms and the old gentlemen who had failed to interest him in mathematics at Cambridge.

Here are some passages from Francis' account of his father's attitude to plants: 'I used to like to hear him admire the beauty of a flower, it was a kind of gratitude to the flower itself, and a personal love for its delicate form and colour. I seem to remember him gently touching a flower he delighted in; it was the same simple admiration that a child might have. He could not help personifying natural things. This feeling came out in abuse as well as praise e.g. of some seedlings—'The little beggars are doing just what I don't want them to.' His emotional attitude to animals was one of profound aesthetic admiration. One of his favourite words was 'wonderful'. Here is a typical passage concerning the second stage larvae of barnacles. 'They have six pairs of beautifully constructed natatory legs, a pair of magnificent compound eyes and extremely complex antennae; but they have a closed and imperfect mouth, and cannot feed.'

In India we expect and find this attitude in saints. But it does not issue, as in Darwin's case, in increased knowledge. The usual effect is a flood of sympathy with animal and human suffering which affects a few thousand people and then degenerates into a new set of ritual prohibitions. It led Darwin to observe the objects of his love with great accuracy. Darwin, then, from the Hindu angle, had some at least of the attributes of a saint.

The movement of art in the last century has been away from Darwin, not only in Europe and North America, but in other countries strongly influenced by them. We are less interested in the details of natural objects than were our grandparents. I expect this is a mere symptom of the senility of 'Western' culture. I do not expect any cultural renaissance until scientific research is an honoured and powerful occupation. When scientists, and particularly biologists, can influence taste, I think the programme for visual art will be 'Back to Albrecht Durer'.

Even before this, I venture to hope that Darwinism may be starting to affect our logic. We do not always realize how much of our ordinary thought is due to Aristotle's difficulties in classifying animals, and to the methods of classifying them which he finally adopted. The greatest Christian theologians, including St Thomas Aquinas and Calvin, adopted Aristotle's logic, though not his metaphysics.

Aristotle's logic is based on similarities. Darwin in the last chapter of the *Origin of Species*, foreshadowed a logic based on differences. I quote two sentences only. 'Systematists will have only to decide (not that this will be easy) whether any form be sufficiently constant and distinct from other forms, to be capable of definition, and if definable whether the difference be sufficiently important to deserve a specific name.' 'Hence, without rejecting the consideration of the present existence of intermediate gradations between any two forms, we shall be led to weigh more carefully and to value higher the actual amount of difference between them.'

We can now see that this last sentence was the programme for a whole branch of statistics. Today we can answer two questions which could not be answered in Darwin's time. First, 'Does population A of animals, plants, or men differ significantly from population B, or could the observed difference be due to randon sampling from the same larger population?' Second, 'Does population C differ more or less from population A than from population B?' What is perhaps of most significance is that the statistical methods devised by Gossett, Mahalanobis, and others to answer these questions as to biological data are now becoming important in physics, geology, and other sciences. Many scientists think that all the sciences will become statistical. If so, Darwin will be recognized as a pioneer in this development.

In the field of ethics Darwinism has probably so far been responsible for more harm than good, as a result of gross misrepresentation, for which, however, he himself bears some responsibility. Darwin was led to the theory of natural selection by reading Malthus. But natural selection operates in a population so fortunate that there is room for every member of it. For example, the human population of North America has been increasing steadily since (AD 1700) or earlier, and for two centuries land was available on the open frontier. However, some of the inhabitants in 1700 left many more descendants than others. There was selection for fertility, resistance to disease, and other characters. The population of France has been nearly stationary for sixty years, but not through famine or pestilence. However, natural selection occurs within it.

Again Darwin naturally concentrated on obviously adaptive characters, such as teeth, horns, and the like, which are of value in the struggle between predator and prey, or between competing males. Their value is more obvious than, say, that of the production of granulocytes which are needed to resist many infections. However, natural selection is far more efficient in eliminating human babies without granulocytes (a recessive character) all of whom die in their first year, than wild mammals whose black colour renders them conspicuous to enemies.

In consequence, although he repeatedly pointed out the importance of physiological adaptation, he certainly left the impression that the struggle for life was analogous to war and economic competition in the human species. And Darwinism was used to justify such activities.

The persons who did so were presumably aware of Jesus' statement 'Blessed are the meek, for they shall inherit the earth.' They were not aware that this statement is substantially true, both in human and evolutionary history. Centripetal selection is normal. That is to say, extremes leave fewer offspring than animals or plants near the average. This appears to be so even if selection is occurring fairly rapidly. Giant species also appear to be much less likely than those of moderate size to leave descendants.

It is a commonplace of human history that ruling classes die out. They may be massacred, but infertility seems a commoner fate. Fisher has argued cogently that the practice of marrying heiresses (who must be members of small families) has concentrated genes for infertility among ruling classes. Kinsey reported sexual behaviour making for low fertility among the richer and better educated Americans. Whatever the reasons economic success is usually correlated with biological failure.

The American negroes offer a conspicuous example of the truth of Jesus' statement. An appreciable fraction of West Africans were sufficiently meek to be capable of living as slaves, which members of prouder races were not. In consequence their descendants are now in a majority in several regions of the American continent and its neighbouring islands.

If this fact of the survival of the meek is ever realized, the consequences may be surprising. I cannot myself foresee them. For the meek do not want to inherit the earth. I have been studying the theory of evolution fairly intensively for some forty years, and I am convinced that, given the facts of genetics, natural selection can be relied on to produce unexpected results.

To sum up, Darwin was too great a man to assess just yet. In each succeeding generation new aspects of his work appear important. Those which I have emphasized may appear less important fifty years hence. But perhaps the perspective of Darwin from an Indian point of view may be a corrective to the 'Western' and Soviet perspectives.

56

Keeping cool

In northern India we had in 1958 the hottest summer for thirty years. The monsoon arrived late, but brought cool air. However, in Bengal we are not getting our usual share of rain, and the weather is getting warm again. Where I work there is usually no serious problem in keeping cool, for we have plenty of good electric fans, and I seldom have to work out of doors between 10 a.m. and 3 p.m. But my wife has to teach a class whose members look at small living insects through microscopes. If fans were turned on, or even if the windows were opened on a windy day, the insects would be blown away. So she and her pupils have to put up with the same conditions as ordinary people.

The problem of keeping cool has been thoroughly studied in England in connexion with the ventilation of mines and factories. It has been little studied outside them. No doubt the Americans claim to have solved the question by air conditioning. But this is very costly, and one feels the heat intensely out of doors on leaving an air conditioned room. It is time that a serious and scientific study were made of this question in India. In England we know the temperatures which people prefer if they have a choice. We also know the temperature at which accidents are fewest in some kinds of factories. For they increase in frequency at higher or lower temperatures than about 13°C. At higher temperatures it seems likely that workers get sleepy and lethargic, at lower temperatures their fingers may get numb and they are more likely to drop things. If workers are given a choice, they usually like to work at a temperature a little above that which reduces accidents to the lowest level. I do not doubt that in India the best temperatures would be higher, except perhaps for very hard work indeed.

How do we keep cool in hot weather? The answsr is not at all simple. Perhaps the first thing of all is to wear sensible clothes. I am writing this in a pair of cotton shorts and no more. Soon, in deference to other workers here, I shall cover up most of my body. I regard this as a needless concession to custom, in either sex. A generation ago women in many parts of South India were not ashamed to show their whole bodies above the waist. And in Europe a mother may feed her baby in public without blame. This is particularly common in France, and thus men are not violently excited by what they see. However, I do not claim the right exercised by the poor and the holy, to work or walk in a dhoti or shorts. But I do slightly pity my male colleagues who wear a cotton vest under their shirt or kurta. And my wife, who wears a very opaque cotton blouse, a black skirt, and little else, pities her Indian colleagues who wind themselves up in a number of layers of cloth. No doubt some of them pity her for wearing black like a Catholic nun, and carrying no jewellery.

In the course of a day a man in India who is not working very hard produces about 2,500 kilocalories of heat, that is to say enough to heat 100 kilograms of water through 25°C. If his heat capacity is equivalent to 50 kilograms of water this means that if he lost no heat at all his temperature would rise about 2°C per hour. So in two hours he would have a pretty severe fever, and in three or four he would be dead. Besides this a man may gain a good deal of heat from the sun which he must get rid of if he is to avoid a rise of temperature. The human body can lose heat in two ways, by warming air and evaporating water. The latter is much more efficient. To get rid of 100 kilocalories per hour one must warm about 270 cubic metres of air through 1°C, whereas only 170 cubic centimetres of water or 4 litres per day need be evaporated.

If the air is cooler than the human body, it can carry away some heat, even if it is so moist that it can take up no more water. Hence a fan or a breeze has a cooling effect. If the air is also dry there may be a lot more cooling from evaporation. If the air is hotter than the human body it will heat it, but if the air is dry so much heat can be lost by evaporation that a man can keep cool. My father, who studied this matter about fifty years ago, remained for 32 minutes in a room where the air temperature was 83°C, or 182°F, but as the air was fairly dry, his temperature only rose slightly. The temperature of a wet bulb thermometer, that is to say a thermometer whose bulb is wrapped in a wet cloth, and cools itself by evaporation, as a man does by sweating, gives a fairly good measure of the effect on a man. If the wet bulb temperature rises above 31°C a man is in danger of death from heat stroke in still air, but in a good air current he can stand 34°C. However, if he is working even moderately these temperatures must be lowered by three or four degrees.

When the air is hot and dry an air current cools him as long as any sweat is left on his skin. But if the air current is strong enough to dry him, any further increase will bring more heat to his body from the air, without any increase in the rate of heat loss. That is why people go indoors if they can during a hot dry wind such as the lu (hot wind) of northern India.

Cows and horses sweat much like men, but dogs, for example, do not. When a dog is hot it produces a lot of very watery saliva, and breathes very quickly. There is enough evaporation from the mouth to keep it cool. But in hot dry air a fan is of no use to a dog, though it would be to a cow. Insects, on the other hand, do not sweat. Unless they have a good supply of water to drink they are killed much more quickly by hot dry air than by moist air at the same temperature. They simply dry up. This is why there are so few during the hot dry weather in India.

I hope that twenty years hence most villages in India will have electric power and that by that time most households will have enough money to buy an electric fan. But before that time I also hope that at least as much research will have been done on keeping cool in India as has been done in Europe. The first subject for such research should be the *kuskus**, as in this part of India we call the wet curtains used to cool air in the dry weather. They cool the air by evaporation. But I am sure that the air currents used with them are

* A grass, *Vetiveria zizanioides*.

usually too weak or too strong to produce the greatest possible cooling effect. I could doubtless frame a theory of cooling by this means, but I know from long and sometimes bitter experience that in such cases theory is no substitute for experiment. Perhaps an equally important topic is the physiology of babies in hot weather. Even in Europe a fair number die from water loss in hot weather. We do not know how far the cooling needs of babies differ from those of adults. In a recent lecture to a medical society I pointed out the need for such research, and the fairly obvious fact that it must be done in India since a European mother would not allow her baby to be exposed to the temperature prevailing in UP in May. A newspaper reporter accused me of demanding experiments on Indian babies. The experiments are done on lakhs of babies every year, and some die in consequence. But as nobody measures the temperature rise, water, and salt loss, and so on, of babies, they die in vain, and more die next year.

There are plenty of other topics to be investigated in connexion with keeping cool in India. None of them demand expensive apparatus. But they cannot be done in a hurry, and they must be done in a number of different parts of the country, for the problems of Madras and Delhi are obviously different, for example the *kuskus* is much more use in Delhi than in Madras. I hope that this article may encourage some of India's young scientists to tackle this important question.

57

Drug-resistant bacteria

I read that some highly placed people in India are drawing attention to the existence of races of bacteria which resist antibiotics, and the dangers of their spread. I am slightly amused when correspondents inform the public that this is a recent discovery, since I was certainly referring to it in popular articles before 1940, and have since written a scientific paper which perhaps makes its investigation a little easier.

The drugs to which resistance develops very frequently are the so called antibiotics such as sulfaguanidine and penicillin. These all seem to act in much the same way. Their molecules are built on nearly the same plan as molecules of an essential foodstuff of the bacteria. They are like enough to unite with the enzymes which transform the foodstuffs into living material, but they cannot be utilized. So if enough of an antibiotic is added, the bacteria behave as if starved of a particular foodstuff. They may be unable to grow, or they may grow abnormally, and die. If, however, we supply a large amount of the substance of which the antibiotic is a dummy, the bacteria can start growing again. This phenomenon is called competitive inhibition. So far as I know it was discovered by Armstrong in 1904, and named by myself in 1930, so it is not very novel.

The action is quite different from that of an ordinary poison such as mercuric chloride, sometimes called 'corrosive sublimate'. This unites with a variety of enzymes, but cannot be crowded off them by adding foodstuffs. Let me try to draw an analogy. If I wished to paralyse the economic life of an Indian city I might fly over it in an aeroplane scattering false one rupee notes, like enough to the genuine ones to deceive the public, but not like enough to deceive bank clerks. This would cause great confusion, and make it hard to use the genuine notes.

If bacteria are subjected to these competitive inhibitors, whether in a human patient or in a culture in a laboratory, new races may arise which are more or less resistant. The most usual, though not the only way, in which this occurs, is the selection of individual bacteria which have varied in such a way as to become resistant. The idea that this happens arouses quite heavy emotional resistance in some biologists, not only in the USSR; 'Variation', they say, 'is a random process, a matter of chance; so you are reducing natural selection to a combination of chance and killing'. The answer to this criticism is that variation is only a matter of chance in that I cannot yet predict that the third seed in an ear of rice will give a plant with green leaves and the fourth one with purple leaves. I can predict that about a quarter of the seedlings will have purple leaves. I cannot predict their arrangement in an ear, and I do not think anyone else will be able to do so for many years to come. What sort of variation will occur is as predictable and intelligible as any other biological event. Sometimes we know enough to predict it very accurately, more usually we can only do so roughly.

The sort of variation which is selected varies considerably. Sometimes the bacteria actually come to require what was a poison to their ancestors, and die when it is no longer supplied to them. More usually they alter their chemical process so as to produce a needed substance by a different method from their ancestors.

However, some bacteria adopt themselves much as a man, by practice, can learn to lift weights which he could not lift at first. One bacterial species can learn to produce an enzyme which destroys penicillin. The learning process has been studied in great detail by Pollock, using penicillin containing radioactive sulphur. This enables us to measure extremely small amounts of it. Each bacterium, for every molecule of penicillin is fixed to it, produces a new molecule of enzyme about once in 90 seconds. If the penicillin is washed away, it makes no more of this enzyme, but what it has made is gradually divided among its progeny. So the capacity to destroy penicillin is lost after ten generations or less. This is rather like what happens in a human society when any branch of art or knowledge becomes purely traditional. One certainly needs tradition, but if one has nothing else, a time comes when posterity copies every attribute of a great man except his greatness.

There is no doubt that in countries such as the USA and England where antibiotics are used on a large scale, drug-resistant bacteria are now much commoner than they were when the drugs were first used. However, it is doubtful if they will replace the original types. For resistance is only achieved by an alteration in chemical make-up which is somewhat wasteful in the absence of the drug, and therefore resistant races can rarely if ever spread in competition with the more sensitive types so long as they are not being selected by the drug. The danger is much greater in a country like America where penicillin is used in a great many unnecessary cases than in one like India where it is very rarely used even when it could save a life. It is less in a country where the medical service is mainly nationalized, as in Britain, than in one where doctors have a financial interest in supplying needless drugs.

Fortunately, perhaps, another danger may help to cut down the needless use of penicillin. So far as I know no case is recorded where a dose of it killed a man or woman who had never had a dose of it before. But it has killed a fair number of people who have had one or more previous injections, and become sensitive to it. So doctors will become more careful of prescribing it unnecessarily.

I do not think we can lay down a correct policy at present. Ideally, when a powerful drug is discovered which will usually destroy the bacteria causing a disease, or better, when several such are found, a worldwide campaign should be made to wipe out the disease for ever. This is not possible until the whole world is economically developed to the stage where it can be done. In India a great and fairly successful effort is being made to make malaria into a rare disease. But it would not be worth trying to abolish it altogether unless this were also being done all over the world. In many African countries, for example, no serious effort has yet been made to prevent malaria. Nor can success be hoped from such an effort until enough Africans have been educated to lead their own peoples in the fight.

I am sure, however, that even if a few private medical practitioners in India use antibiotics when they should not be used, the country as a whole needs enormously more than are available. Certainly we should accept 'aid' in the form of such drugs from any foreign state which will let us have them, while I fully agree with the policy of our Government in refusing to accept weapons which may only 'aid' armies to assume unconstitutional power. But we must make our own antibiotics, and I hope that our third five year plan will provide for larger increase of factories for this purpose than did our second. The danger of excessive use of these substances, which may be a real one in the USA, is still very remote in India. And I hope that by the time it is a real danger our medical service will have been nationalized at least to the extent that this has been done in Britain. As I pay a weekly subscription to the Medical Unit of the Indian Statistical Institute, for which I get most satisfactory service, I am not recommending to others a course of action which I do not take myself.

GLOSSARY

Astika and Nastika technical terms in Hinduism used to classify philosophical schools and persons, according to whether they accept the authority of the Vedas.

Atman or *Atma* is a philosophical term used within Hinduism and Vedanta to identify the soul.

Gayatri Mantra the holiest verse of Hinduism.

Gita the *Bhagavad Gita* (Song of God) is a Sanskrit text from the Mahabharata epic, comprising 700 verses.

Kaliyuga one of the four stages of development that the world goes through as part of the cycle of Yugas, as described in Hindu scriptures.

Kapila a Vedic sage, author of basic principles of the Samkhya system of Indian philosophy.

Krishna deity worshiped across many traditions of Hinduism; an incarnation of Vishnu.

Kshattriya warrior caste.

Kurukshetra (Battle of), a city in the present-day Indian State of Haryana.

Narikel coconut.

Parasurama sixth of the ten incarnations of the Hindu god Vishnu.

Prakriti according to the Vedas, the basic matter of which the Universe Consists.

Rama the seventh Avatar of Vishnu, Rama is one of the most popular deities in Hinduism.

Rishis sages of ancient India.

Samkhya it is one of the six schools of classical Indian philosophy, founded by the sage Kapila. It is regarded as the oldest of the philosophical systems in India.

Siva or **Shiva** one of the principal deities of Hinduism.

Svarga heaven.

Tapas in Hinduism, ascetic practice voluntarily carried out to achieve spiritual power or purification.

Upanishads part of the Hindu scriptures. They form the core spiritual thought of Vedanta.

Vedanta a spiritual tradition within Hinduism that is concerned with the self-realization by which one understands the ultimate nature of reality.

Vishnu supreme being of Hinduism, responsible for the maintenance or 'preservation' of the Universe.

Visvamitra one of the most venerated sages of ancient India.

Yudhisthira the principal protagonist of the Kurukshetra War and the eldest son of King Pandu and Queen Kunti.

ESSAY REFERENCES

Part I: Essays from the *Daily Worker* (London):

1. How to write a popular scientific article (1941) [excellent advice to writers of all ages]
2. Why I am a materialist [written for the Marxists in 1940]
3. Adventures of words (1944) [Marxists's view of history]
4. What I require from life (1940)
5. What scientists in Russia are doing (1945)
6. How we can date the past (1948)
7. Averages (1945)
8. Quantity and quality (1946) [based on his experiments in physiology using himself as his own guinea pig]
9. Biology and town-planning (1940) [physiology, ventilation, public health]
10. Inventions that made men free (1947) [ancient Greco-Roman inventions]
11. Astronomy (1945)
12. The Milky Way (1948)
13. Is there life on the planets? (1940)
14. Common sense about the planets (1947)
15. Weather (1940)
16. Frost (1948)
17. Is man a machine? (1947) [dialectical approach to men and machines]
18. Breathing (1945) [Haldane's research in submarine physiology]
19. Why are you left-handed? (1948) [evolution]
20. Domestic animals (1947) [origin of domestication of plants and animals]
21. Overcrowding at the zoo (1949) [humane captivity of animals in zoos]
22. Life slows down for winter (1945) [physiological adaptation in cold temperature]
23. Bird migration (1948) [research project for the laymen]
24. Why the robin sings (1950)
25. How bees communicate (1949)
26. The mysterious eels (1939)
27. Movies for toads (1951) [experiments in toad behaviour]
28. The chicken or the egg? (1948) [Marxist view of evolution]
29. The mathematics of evolution (1950) [evolution in human populations]

30. Back to the water (1948) [animal evolution and human societies]
31. The common cold (1947) [research on common cold]
32. Medical measurements (1945) [monitoring health]
33. The differential calculus (1944) [a primer]
34. What is the fourth dimension? (1948) [measuring space]
35. Relativity (1948)
36. Matter and energy (1945)
37. How we measured the atom (1946)

Part II: Essays from the Indian Period (1957–64), from the Hindu, Chennai, India

38. Some statistical adventures (1957) [statistics in physiology and biology]
39. Some reflections on non-violence (1959) [non-violent biological research]
40. Science and floods (1959)
41. Colliery explosion (1959) [safety in mines]
42. Deep mines (1959) [mining methods]
43. The sound of one hand (1959) [Buddhism and science]
44. Hitting the moon (1960)
45. The dog in the sputnik (1957) [first dog in space; Soviet space research]
46. What I want to know about Gagarin (1961) [first man in space and physiology]
47. Some autumn stars (1958)
48. The Pleiades and Orion (1958)
49. Some neighbouring stars (1958)
50. The seven rishis (1958)
51. Jyestha (1959) [binary star Antares, constellation Alpha Scorpius]
52. Simplifying astronomy (1959)
53. New light on memory (1958) [relation between mind and brain]
54. Relations between biology and other sciences (1952) [address to the Indian Science Congress]
55. Darwin in Indian perspective (1958) [Christian and Hindu views of Darwin's theory]
56. Keeping cool (1958) [physiology of keeping cool in a hot climate]
57. Drug-resistant bacteria (1958)

INDEX